GOLD AND SILVER
LEACHING, RECOVERY AND ECONOMICS

PROCEEDINGS FROM THE 110th AIME MEETING

CHICAGO, ILLINOIS, FEBRUARY 22–26, 1981

EDITORS

W.J. SCHLITT W.C. LARSON J.B. HISKEY

SOCIETY OF MINING ENGINEERS AIME

SPONSORED BY

SOLUTION MINING COMMITTEE
MINING AND EXPLORATION DIVISION
SOCIETY OF MINING ENGINEERS OF AIME

Printed in the United States of America
by Lucas-Guinn Co., Hoboken, New Jersey

Library of Congress Catalog Card Number 81-68558
ISBN 0-89520-289-1

PREFACE

Gold and silver are the latest commodities featured by the Solution Mining Committee in its efforts to develop technical programs which focus on topics of current interest in the broad field of solution mining. These programs have taken the form of minisymposiums devoted to in-depth assessments of the selected subjects. Past efforts have included the 1979 New Orleans symposium on in situ uranium leaching and ground water restoration and the 1980 Las Vegas symposium on leaching and recovering copper from as-mined materials. Like the earlier ones, the 1981 symposium on leaching and recovering gold and silver was held as part of the AIME annual meeting, this time in Chicago, Illinois.

Gold and silver proved to be a particularly attractive topic since the program was held at a time characterized by historically high precious metal prices. These prices have resulted in a worldwide boom in precious metal mining. In addition to developing new mines, the activities have included reopening old operations, expanding existing plants and processing previously discarded wastes. While continued high prices have brought about an expansion in exploration and mining activities, the high cost of capital has made plants more expensive to build and inflation has driven up operating costs substantially. When combined with declining ore grades and more difficult mining conditions, technical innovations are needed to maintain the viability of many operations. These innovations are addressed in the first 12 chapters of the book.* The first chapter provides an overview of current leach operations while the next two examine the treatment of problem ores (carbonaceous, sulfidic and manganiferous). The next group of papers deals with heap leaching, an approach now gaining favor due to its low cost, rapid implementation and relative simplicity. These are followed by a paper on combined flotation and leaching. All these commercial leach operations utilize cyanide lixiviants and the next two chapters deal with treatment of such solutions to recover the values and avoid discharge of hazardous wastes. Use of thiourea as an alternative to cyanide is discussed in the following papers while assay methods and problems are covered in the twelfth paper.

As pointed out above, technology cannot be divorced from economics. For this reason, Leons Kovisars and Nellie E. Guernsey of the SME Minerals Resource Management Committee developed a complimentary session dealing with the economics of precious metals. Substantive contributors by bankers, consultants and commodity traders have been included in this volume. These provide the link between technology and the market. Therefore, inclusion of the papers on econimics makes the volume on gold and silver much more comprehensive. As the organizers of the session on economics point out, the market determines the price while geology and metallurgy provide the basis for production. If the price is right, miners will make enough gold and silver available to meet the demand.

The Editors May 22, 1981

---

* Not all these papers could be accommodated in the solution mining sessions. Some were actually presented in other Society of Mining Engineers or Metallurgical Society sessions.

CONTRIBUTING COMMITTEES

SOLUTION MINING

William C. Larson, Chairman

J. Brent Hiskey, Program Planning

Daryl R. Tweeton, Publications

W. Joseph Schlitt, Past Chairman

MINERALS RESOURCE MANAGEMENT

H. Lyn Bourne, Chairman

Nellie E. Guernsey, Vice Chairman

Leons Kovisars, Program

The committes are most appreciative of the time and effort contributed by the authors in preparing their presentations and manuscripts. The committees would also like to thank Myrna R. Anderson, Fred C. Dunford and Donna Sweatfield of Kennecott Minerals Company; Carol Walters of Mineral Systems Incorporated and Marianne Snedeker of the Society of Mining Engineers for their efforts and support in preparing this volume.

TABLE OF CONTENTS

Chapter 1

THE STATUS OF GOLD AND SILVER LEACHING OPERATIONS IN THE UNITED STATES

Peter G. Chamberlain
Michael G. Pojar
U. S. Department of the Interior
Bureau of Mines
5629 Minnehaha Avenue South
Minneapolis, Minnesota 55417

ABSTRACT

Solution mining, primarily by heap leaching, has emerged as an important method for mining gold and silver in the United States. Heap leaching offers many economic and environmental advantages over conventional production techniques, in particular for lean ores and old mine waste rock. In order to more effectively pursue research on heap, dump, and in situ solution mining of gold and silver, the Federal Bureau of Mines conducted an extensive survey of leaching practices and problems. This presentation summarizes trends and statistics. Leaching is projected to provide 20 pct of primary mine produced gold and 10 pct of primary silver by the end of 1982. Current leaching practices in the United States are reviewed; operational problems and research programs to alleviate those problems are discussed.

Introduction

As gold and silver prices have dramatically risen in recent years, many new mining ventures have opened in western United States. Unlike many other heavy metals, gold and silver can be profitably mined in small operations because of their high unit values. One mining technique that seems ideally suited for small deposits of gold/silver and for certain large, low-grade oxidized "disseminated" deposits is solution mining (referred to as leaching hereafter). Leaching requires lower capital investment and operating costs than conventional mining and milling, has a faster start-up time, and is generally considered to be safer and less degrading to the environment. The major trade-off is the lower or uncertain recovery associated with most leaching operations.

The Bureau has conducted various research projects to improve gold and silver leaching practices. Current leaching practices and engineering data from gold and silver leaching operations have also been compiled by the Bureau to assist potential new leaching operators in evaluating leaching possibilities and establishing reasonable operating parameters (Chamberlain and Pojar, 1981). This presentation is a continuation of efforts to provide the mining industry with up-to-date information on gold and silver leaching techniques and production statistics. Specifically, the purpose of this presentation is to briefly review the state-of-the-art in leaching practices, highlighting operating problems and research being performed to resolve them, and to provide recent production statistics that illustrate the role of leaching in gold/silver mining.

Although gold and silver have been recovered by cyanidation following conventional milling since the late 1800's, solution mining or leaching as discussed here has gained momentum in the past decade.

There are basically three new types of leaching systems used today--heap, dump, and in situ ("vat" leaching accompanying conventional milling operations will not be considered). If ore is mined or gathered from old mine waste rock piles and hauled to specially prepared impervious pads, the method is termed "heap" leaching (Figure 1). The rock is frequently but not always crushed before being placed on the pad. Leaching solutions are applied to the top of the heap, usually through fixed or oscillating sprinklers, and allowed to percolate down through it. The solution containing dissolved values eventually reaches the impervious pad and flows along it to a collection pond. From the pond the solution is circulated through the recovery plant where the mineral value is removed by zinc precipitation or adsorption on activated carbon (charcoal).

If old mine waste rock piles or dumps are judged to contain sufficient mineral value to justify leaching and the solutions can be controlled without appreciable losses, the pile is "dump" leached as is (Figure 2). Dump leaching is essentially the same as heap leaching except that the rock is not moved to a specially prepared pad. Since an old dump has no impervious pad, the pregnant leaching solutions must move to a collection pond along the ground/dump interface after passing down through the dump. Although dump leaching is common in copper mines, the only gold and silver operation indicating dump leaching was Carlin's Bootstrap mine.

Finally if the ore is broken and left in place or if it is conducive to proper fluid flow without blasting, the ore can be leached "in situ" or in-place (Figure 3). An exposed orebody can be leached in situ by spraying solution on the surface and collecting it in recovery wells after it has percolated down through the ore. For buried orebodies, the solution must be injected into the formation via injection wells and recovered from adjacent recovery wells. Although copper and uranium have been leached in situ, there have been only sporadic attempts to develop in situ gold and silver leaching operations; none have been commercial.

Gold/Silver Leaching Statistics

The role of gold/silver solution mining is best determined through an examination of production statistics and engineering data. The first statistic is one that we see so often in our overall mineral industry--import reliance. As with many critical minerals, a high percentage of gold and silver must be imported from other countries. Figure 4 shows that in 1979 we imported 56 pct of the gold consumed and 45 pct of the silver (Bureau of Mines, 1979). The market for both metals is extremely complex and will not be analyzed in this presentation. Suffice it to say that both carry strategic importance as well as their monetary exchange value and collector's/jewelry value. The importance of this statistic is that it is in the national interest to promote orderly growth in gold and silver mining. Solution mining is a method that lends itself well to stimulating domestic production because it is better suited for smaller or lower grade deposits than is conventional mining/milling. Lower capital requirements and shorter lead time to establish a leaching operation, have made the method especially attractive for new and/or small operators.

Figure 5 shows that in addition to imports, production of gold and silver from scrap is a significant factor in providing these metals for domestic consumption (Bureau of Mines, 1980). While this may point out the inherent difficulties in making an immediate impact on our supply of gold and silver through innovations in any one mining method such as solution mining, it again indicates the need for expanding the role of primary mine production in supplying gold and silver. And the high demand for gold and silver certainly provides a suitable climate for expanding solution mining activities.

Since this paper is concerned principally with primary or mine production, Figure 6 shows that about 40 pct of the gold is produced as a byproduct of mines targeted for other commodities (Bureau of Mines, 1980). This means that a large portion of our gold supply is dependent on mining technology and market forces developed for other commodities, and this has become more prevalent with each passing year. Although the price of gold may influence the economics of these mining operations (and in fact gold may even replace the other commodity as the primary target), in general this portion of the supply is relatively insensitive to normal supply/demand relationships.

The situation is even more pronounced for silver. Roughly two-thirds of the silver mined in this country occurs as a byproduct of other metallic ores. Lead/zinc ores and copper ores are the chief sources of byproduct silver (Bureau of Mines, 1980).

According to the most recent figures obtained by the Bureau of Mines, a total of 51,037 troy ounces of gold and 28,949 troy ounces of silver were reported as being produced by heap leaching crude gold/silver ores in 1979 (Lucas, 1981). This amounts to 5 pct of total gold and much less than 0.1 pct of the total silver produced in the U.S.A. Since all solution mining production is not reported, per se, to the Bureau's mineral industry survey team, the authors estimate the amount of gold and silver produced by solution mining based upon engineering data gathered from site visits and numbers available in the literature. These calculations indicated that about 109,000 ounces of gold and 131,000 ounces of silver were produced by solution mining in 1979. If the unreported total production was the same as was the unreported leaching production, the corresponding percentages would be about 10.6 pct of the gold and 0.3 pct of the silver (Figure 7). Gold and silver produced by leaching copper ores are not included in these figures.

It is clear that leaching has already achieved prominence in producing gold but is not a significant factor for silver. By comparison, an estimated 9 pct of uranium was produced in 1979 by in situ solution mining techniques (Larson, 1980) and from 10 to 16 pct of copper mined in Arizona is produced by heap/dump leaching (Schlitt, 1980).

Although solution mining or heap leaching data have not been compiled along with other mining statistics for publication in the Minerals Yearbook series, I feel that this production has remained fairly constant for several years. This is primarily caused by the heavy influence of one or two mines which produce the majority of gold and silver.

All indications are that the future trend will be towards considerably increased solution mining. The advantages previously mentioned, coupled with the continued strong gold and silver market, have led to much interest in solution mining. From all reports of mining industry activities in establishing operations, solution mining is experiencing an exciting growth surge. Judging from operations that came on line in 1980 or are planned for 1981 (and expansions of current operations) by 1982 operators could be producing 250,000 ounces of gold per year by heap leaching which would represent 20 to 25 pct of our domestic mine production (Figure 8).

Silver leaching will also show a positive growth surge in spite of its complex mineralogy and heavy production as a byproduct. Primarily because of the influence of one or two large mines entering leaching production in 1980 or 1981, annual production from solution mining could reach 4 million ounces of silver by the end of 1982. This would represent some 10 pct of the total production (Figure 8).

Production Data

Where does the leaching production come from? Leaching operations functioning in or targeted for gold and/or silver ores are concentrated in western United States, along a broad belt coinciding with mountain ranges that historically have hosted the bulk of the country's precious metal mining activity (Figure 9). Within this belt, the density of operations is affected by many geological and topographical factors as well as the difficulty in obtaining permits to leach. A thorough discussion of these factors is beyond the scope of the presentation. Nevada and Arizona have provided most of the leaching operations to date.

A detailed presentation of production and engineering data for gold/silver solution mining operations is being prepared by the Bureau of Mines (Chamberlain and Pojar, 1981). The following engineering data are all taken from that report. Table 1 shows that there are now more gold-oriented leaching operations than silver. Although gold and silver commonly occur together it seems apparent that gold targets have attracted more attention than silver. This could be because the chemistry of more gold ores is advantageous for leaching. Many common complex silver ores are difficult to leach whereas gold generally occurs in ores as native gold which is readily leached with cyanide.

TABLE 1. - Target Commodities Being Leached

Primary Commodity	Number of Operations	Value Produced[1]
Gold	16	$33 million
Gold/Silver	7	N.A.
Silver	6	$1.3 million

[1]Based on 1979 average prices of $300 for gold and $10 for silver which was determined from total metal produced and its total value.

Value of gold produced in 1979 by leaching compared with that of silver produced by leaching is also shown in Table 1. These figures are based on my estimates of production by leaching and an assumed gold value of $300/ounce and silver value of $10/ounce.

Many operators are "getting their feet wet" by treating old waste rock dumps. Without any front-end mining costs, the operator only needs to provide a leaching solution distribution/recovery system plus an extraction system to leach old mine waste rock. The cash flow generated by leaching this material can then be used to finance development of virgin ore for additional leaching if economics dictate. Of the solution mining operations for which most data have been obtained, eight are leaching waste rock and seventeen are leaching virgin ore. The waste rock being leached originates predominantly from old underground "vein" mines while ore for leaching is mostly mined from surface mines. The ounces of gold and silver are even more heavily weighted towards production from fresh ore leaching operations.

Engineering data from gold and silver leaching operations show some interesting trends, for instance, the lack of dump leaching operations and total absence of in situ leaching operations at this time. The move toward heap leaching reflects the strong desire to control the leaching solutions from an environmental and economical standpoint. This control is usually unavailable unless the rock is placed on a special impermeable pad. A second benefit of heap leaching occurs if the rock is crushed before being placed on the pad because leaching rates and total recovery are significantly higher for smaller particles.

Although uranium and copper have been successfully leached in situ, there are no commercial scale gold/silver in situ solution mining operations. The only recent attempt to leach gold or silver in place is one that was conducted in an old underground vein mine. As shown in Figure 10, water was injected under pressure into a fissure and collected down dip in a trough built in a "strike" drift. The experimenters concluded that the drift and trough were not wide enough and that excessive solution was lost when it bypassed the collection system. The experiment was not concluded because laboratory tests never resolved the chemistry of leaching that particular ore.

Much discussion in recent years has centered around a solution distribution system for leaching solutions. Almost all operators feel that in order to maintain a high dissolved oxygen content in the solution (which is necessary for the leaching reaction), some sort of spraying method is best. Only one operator uses a ponding technique to distribute leaching solution on heaps. Seventeen operations employ oscillating ("rainbird") sprinklers whereas only two each use fixed sprays and Bagdad wigglers. Bagdad wigglers generally require less maintenance than oscillating sprinklers and are more popular among the large operations, but sprinklers have a broader coverage. Fixed sprays are the cheapest but provide smaller and extremely uneven solution distribution. A valuable research project would be to determine the amount of dissolved oxygen required for the leaching reactions and to determine the optimum solution distribution method for achieving the desired amount.

Another controversy has brewed over the relative benefits of recovering the gold and silver by precipitation with zinc (Merrill-Crowe process) or by adsorption onto activated carbon (charcoal) with charcoal adsorption enjoying a slight edge in popularity. Several factors influence the selection. Ores with high silver content appear to dictate recovery by zinc precipitation because carbon does not adsorb silver as cost effectively as it does gold. Zinc precipitation systems are easy to install but require very clean solutions (which necessitates good filtration) and absence of oxygen (which necessitates deareation by vacuum and/or baffles) both of which contribute to capital costs. Economics of charcoal adsorption relate to initial investment in charcoal plus any replacement of it along with costs of stripping and regeneration units.

Problems/Research

Mineralogy

Before the full benefits of leaching can be realized, there are problems that must be solved through diligent research efforts by government and industry alike. Although the acquisition of property and permits may present the biggest headaches to a would-be operator who has discovered an orebody, a discussion of them is beyond the scope of this presentation. Once these obstacles are overcome, an early problem that confronts the would-be operator is that of mineralogy. Does the gold/silver occur in a readily leachable form? Although gold is usually in a free or native state which is readily leachable, silver often occurs in minerals that do not easily leach (Dorr and Bosqui, 1950, and Leaver, Woolf and Karchmer, 1931). For example, if manganese occurs with silver ores, upon weathering its higher order oxidation products can form refractory minerals of silver and manganese (Scheiner, et al, 1973, and Clevenger and Caron, 1925). Potential solutions to this problem are discussed in another chapter of this symposium volume (Chase, 1981).

Deposits that contain organic carbon are not suitable for leaching because the carbon adsorbs the dissolved metal before the leach solutions are recovered. Other refractory ores are those in which the gold or silver is totally encased by an impervious matrix material such as quartz so that the leaching solutions cannot contact the metal. Copper, cobalt, and zinc in the ore may preferentially take the place of gold and silver in the leaching reaction, greatly reducing the reaction with the desired metal. Some ores such as the tellurides or those containing arsenopyrite or antimony must be roasted before they can be leached with cyanide and so do not lend themselves to cyanide leaching (Dorr and Bosqui, 1950). While other leaching solutions have been investigated for use with telluride deposits (for example chloride leaching) no commercial operations have resulted.

Pyrrhotite is a common refractory mineral. Decomposition of pyrrhotite creates ferrocyanide, which removes free cyanide from solution preventing its reaction with the gold or silver (Dorr and Bosqui, 1950, and Hedley and Tabachnick, 1968). Pyrrhotite decomposition also removes oxygen from the solution which greatly decreases the solution reaction with gold and silver.

Research to resolve problems in the chemistry of cyanidation caused by mineralogical setting in ore deposits is continually in progress. Almost every deposit contains quirks that must be resolved by testing in a laboratory. Such research would fill many papers such as this and, hence will not be discussed.

Clay Content

Clay in the ore poses a major problem in leaching operations. The clay particles block leaching solution flow through the ore leaving large portions totally isolated from the leaching solution. Some heaps may contain so much clay that the solution perches on top of the heap with negligible downward percolation (Figure 11).

Research has been conducted by the Bureau of Mines to reduce the nonuniform and poor solution flow through the heaps due to clay particles in the heap (Heinen, McClelland and Lindstrom, 1979 and McClelland and Hill, 1981). By mixing Portland cement with the fine particles and adding water, the mixture will agglomerate into small lumps which remain intact as the cement cures. Flow rates through heaps and recovery rates can be an order of magnitude greater for agglomerated ore over that for normal heaps.

Dissolved Oxygen/Solution Loss

Nonmineralogical problems abound in leaching operations. The requirement for oxygen in the dissolution of gold and silver in cyanide has generally resulted in spraying solutions on the surface of the heap or dump. (The most notable exception being the successful solution-flooding system at the Windfall Mines in Nevada.) Spraying, while increasing dissolved oxygen content, creates rather large solution losses--frequently running 15 to 25 pct--because of the arid climate present at most leaching sites.

A possible research project would be to identify necessary levels of dissolved oxygen and to determine the best means of applying solutions to achieve that oxygen content.

Lime Build-up

Large amounts of lime used to maintain the desired high pH in leaching solutions also cause a probelm by clogging the spray system. The lime eventually coats the distribution system, most critically within the sprinkler heads until solution flow through the heads is severely restricted. At this point the heads must be removed and soaked in hydrochloric acid. This is both time consuming and detrimental to the longevity of sprinkler heads.

The problem of lime build-up in sprinklers has been grappled with from several different directions. Many operators have experimented with and converted to the use of NaOH instead of lime to maintain the desired solution pH. Although NaOH is more expensive, the coating builds up inside the sprinklers and lines at a much slower rate. Other operators have experimented with additives such as Baroid's Surflo-H35* to prevent unwanted precipitation of calcite-type minerals with mixed results.

The Bagdad wigglers first used to distribute solutions at the Bagdad copper mine have been tried

*Reference to specific brand names is made for identification purposes only and does not imply endorsement by the Bureau of Mines.

successfully at several gold leaching operations. Simple to construct, from 1/4 inch surgical rubber tubing (Figure 12), these sprinklers are easy to free from lime build-up simply by stretching and snapping the rubber tubing. The large tubing inside diameter will remain open much longer that the small orifices found in oscillating sprinklers. Finally, one operator reported success in distributing solutions with minimal lime problems with a small, gardening-tractor shaped traveling sprinkler after its orifice was drilled out to about 1/8 inch. Since this sprinkler was not observed in action, its potential will not be commented upon.

Temperature

Cold temperatures severely reduce leaching solution effectiveness. Below 50° F the solubility of gold and silver is significantly retarded. In very cold weather the solutions freeze on the top of the heap instead of percolating down through it. Since most of the current leaching operations are located in mountainous regions which experience cold temperatures during winter, virtually all leaching operations must face this problem. Most shut down for several months in the winter.

Although the problem of cold temperatures has plagued operators for years, surprisingly little research has been done on ways to circumvent it. One notable effort has been the installation of a submersible kerosene heater in the solution holding pond at Smoky Valley's Round Mountain operation. This unit has been successful in modestly warming solution temperatures which increases recovery per unit time and extends the number of days during the year when solutions can be applied (Lefler, 1981).

A research endeavor with excellent potential would be to develop a solar heating unit for heating solutions. Since most leaching operations are established in the western United States where sunlight and areal space prevail, the solar units should be ideal. The rising price of fossil fuels certainly should provide an incentive. Brine evaporating ponds also offer possibilities.

In Situ Leaching

A discussion of research in gold/silver leaching research would not be complete without mentioning in situ leaching. The Bureau has been intrigued with the possibilities of leaching gold and silver in place because of the interesting economic and environmental benefits that may accrue if conditions are right for it. Without delving into a detailed discussion of leaching conditions, suffice it to say that an operation could only be established where reasonable guarantees can be provided that no potable water supply would be affected. The Bureau funded a contract with Mountain States Research and Development in Tucson, to evaluate the feasibility of in situ leaching various commodities. In situ gold/silver leaching appeared attractive especially for the low-grade, oxidized finely disseminated ores. A conceptual design for establishing a leaching operation at a typical site of that type is being prepared by the contractor for release soon.

Summary

Our reliance on imports for both gold and silver is a simple but effective testimony on the need for increased domestic production of these commodities. Solution mining offers unique possibilities for increasing domestic production particularly from low-grade ore and old mine waste for which production would not otherwise be feasible. All indicators point out that the solution mining industry is in the midst of an exciting growth spurt such that it will produce 20 pct of the gold and 10 pct of the silver mined in this country in 1982. Several problems that tend to thwart solution mining operations and research efforts that seek to overcome these problems were identified.

References

1. Bureau of Mines, 1979, "Status of the Mineral Industries," Bureau of Mines Publications Distribution Branch, Pittsburgh, PA.
2. Bureau of Mines, 1980, "Mineral Commodity Summaries 1980," Bureau of Mines Publications Distribution Branch, Pittsburgh, PA.
3. Chamberlain, Peter C., and Pojar, Michael, 1981, "Gold and Silver Leaching Practices in the United States," Bureau of Mines Information Circular, in preparation.
4. Chase, C., 1981, "Improved Recovery of Silver from Manganiferous Ores," (published in this volume).
5. Clevenger, G. H., and Caron, M. H., 1925, "The Treatment of Manganese - Silver Ores," Bureau of Mines Bulletin 226.
6. Dorr, J.V.N., and Bosqui, F. L., 1950, "Cyanidation and Concentration of Gold and Silver Ores," McGraw-Hill, New York.
7. Hedley, N., and Tabachnick, H., 1968, "Cyanamid Wants You to Know: About the Chemistry of Cyanidation," American Cyanamid Company Report, Part II, Article 23, Wayne, NJ, pp 21-72.
8. Heinen, H. J., McClelland, G. E., and Lindstrom, R. E., 1979, "Enhancing Percolation Rates in Heap Leaching of Gold-Silver Ores," Bureau of Mines Report of Investigations 8388.
9. Larson, W. C., 1980, "Uranium In Situ Leach Mining - Environmental Controls," Proceedings of 1st Inter. Conf. on Uranium Mine Waste Disposal, Society of Mining Engineers of AIME, pp 174-192.
10. Leaver, E. S., Woolf, J. A. and Karchmer, N. K., 1931, "Oxygen as an Aid in the Dissolution of Silver in Cyanide," Bureau of Mines Report of Investigations 3064.
11. Lefler, C. A., 1981, "Leaching Practices at Round Mountain," (published in this volume).
12. Lucas, J., 1981, Personal communication of data that will appear in "Minerals Yearbook, vol. I," to be published by the Bureau of Mines.
13. McClelland, G., Hill, S. D., 1981, "Heap Leaching Gold-Silver Ores with Poor Percolation Characteristics," (published in this volume).
14. Scheiner, B. J., Pool, D. L., Sjoberg, J. J., and Lindstrom, R. E., 1973, "Extraction of Silver from Refractory Ores," Bureau of Mines Report of Investigations 7736.
15. Schlitt, W. J., 1980, "Current Status of Copper Leaching and Recovery in the U. S. Copper Industry," Chapt. I in _Leaching and Recovering Copper from As-Mined Materials_, edited by W. J. Schlitt. Society of Mining Engineers of AIME, pp 1-7.

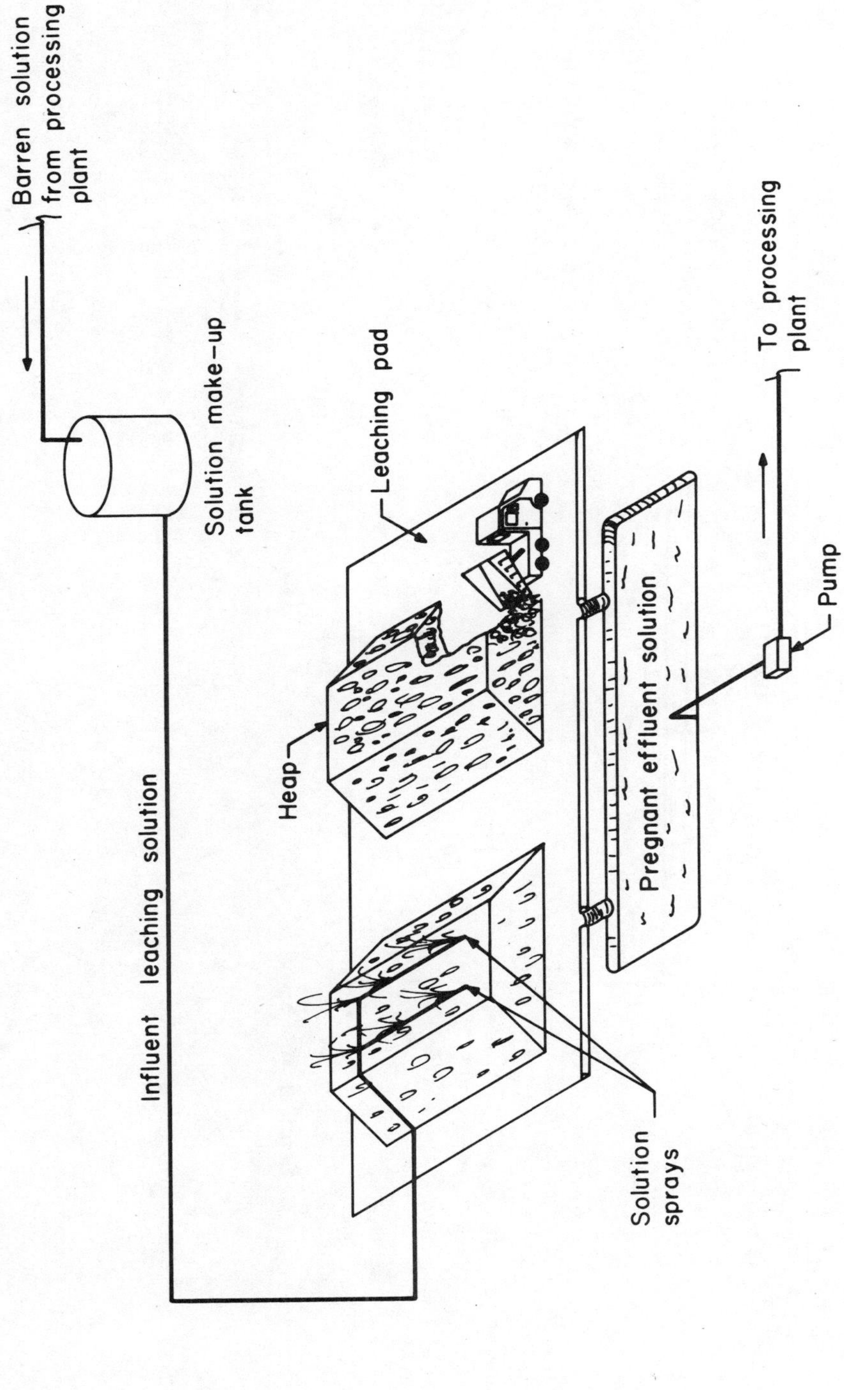

Figure 1. Heap leaching system (Chamberlain and Pojar, 1981).

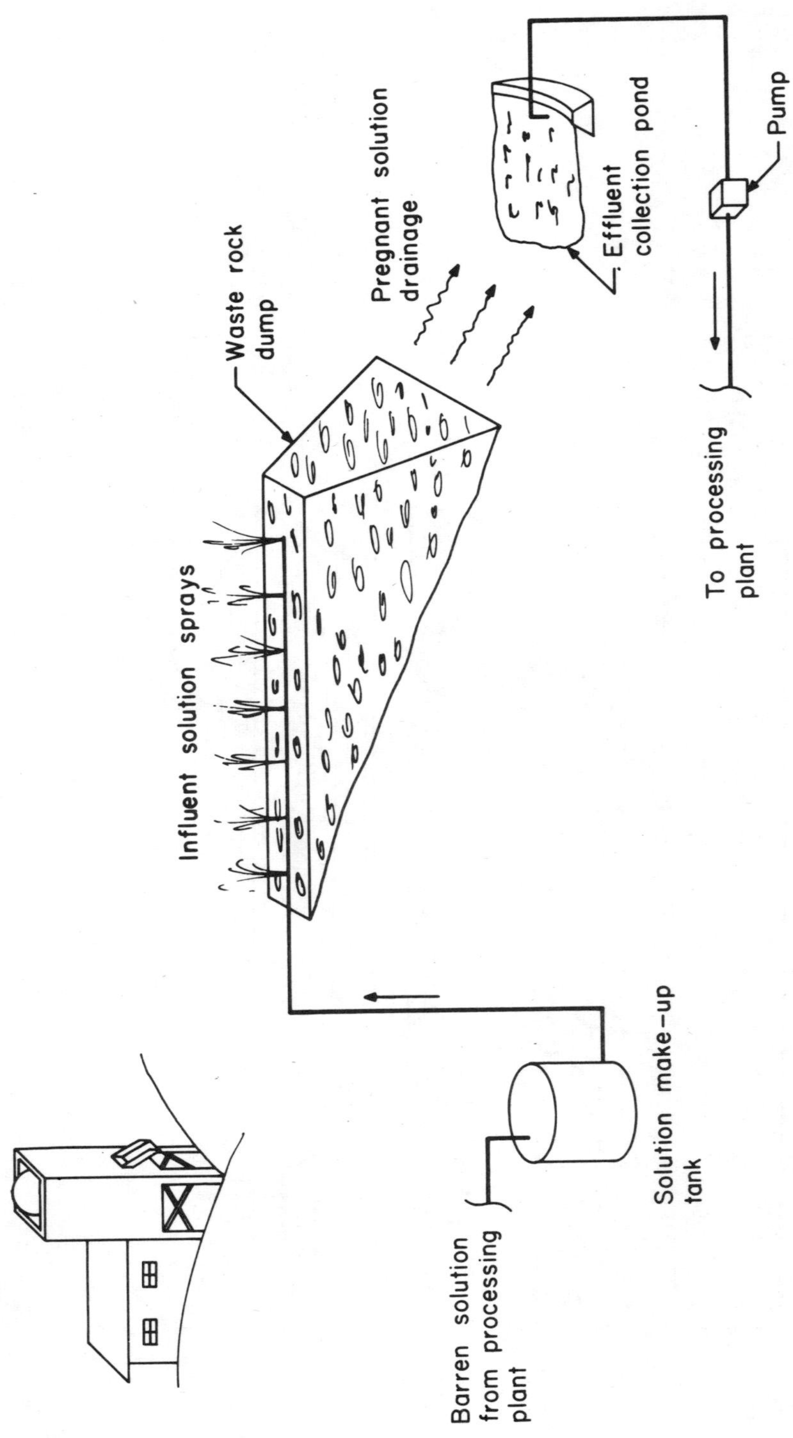

Figure 2. Dump leaching system
(Chamberlain and Pojar, 1981).

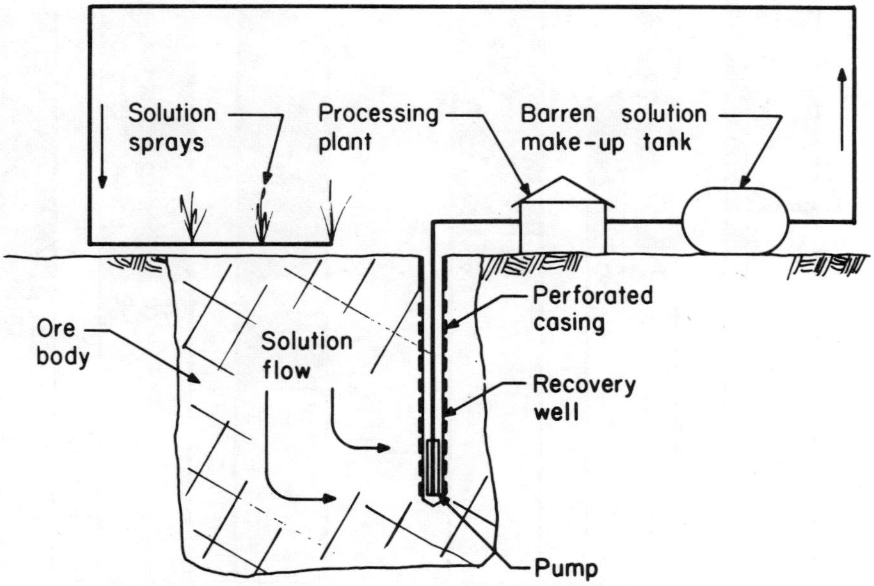

Exposed Ore Body

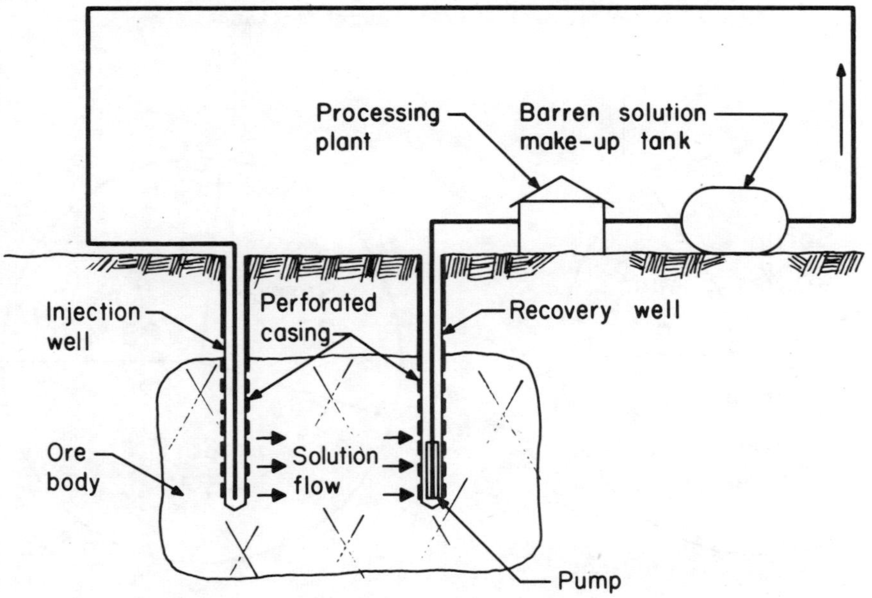

Buried Ore Body

Figure 3. In situ leaching system
(Chamberlain and Pojar, 1981).

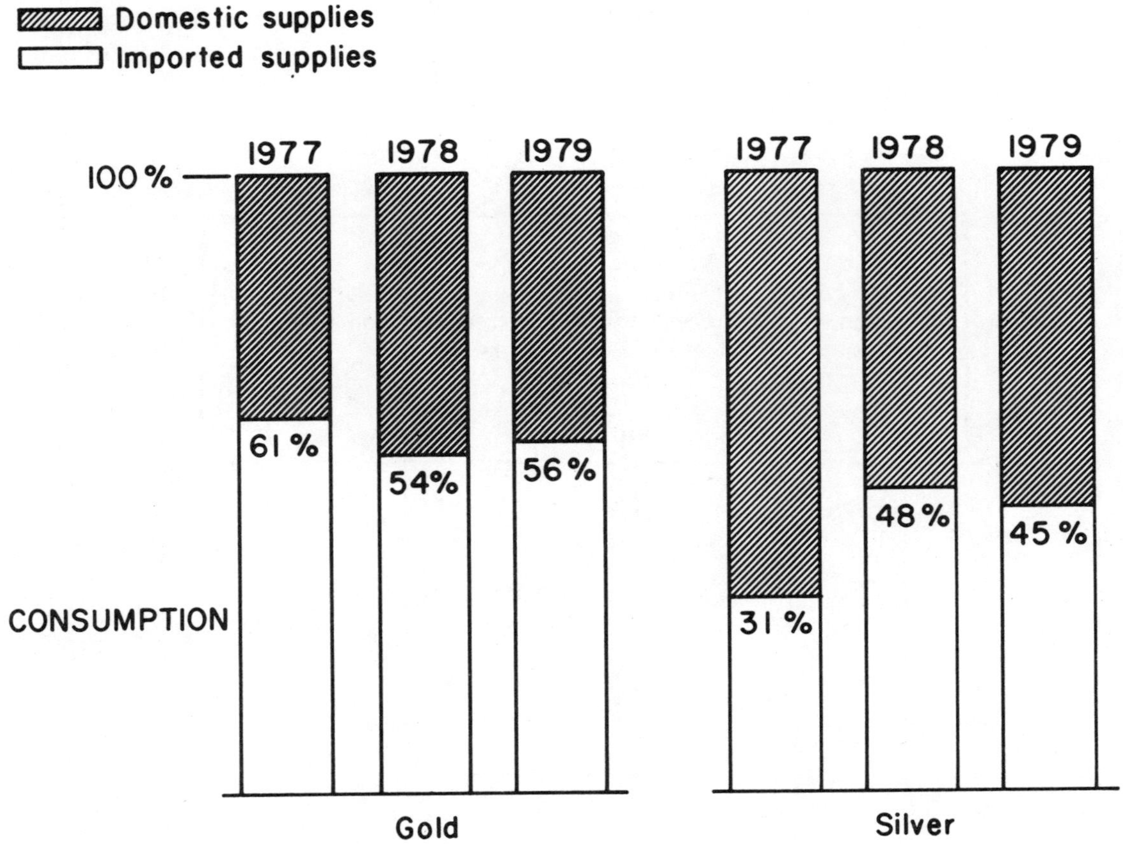

Figure 4. Gold and silver import reliance.

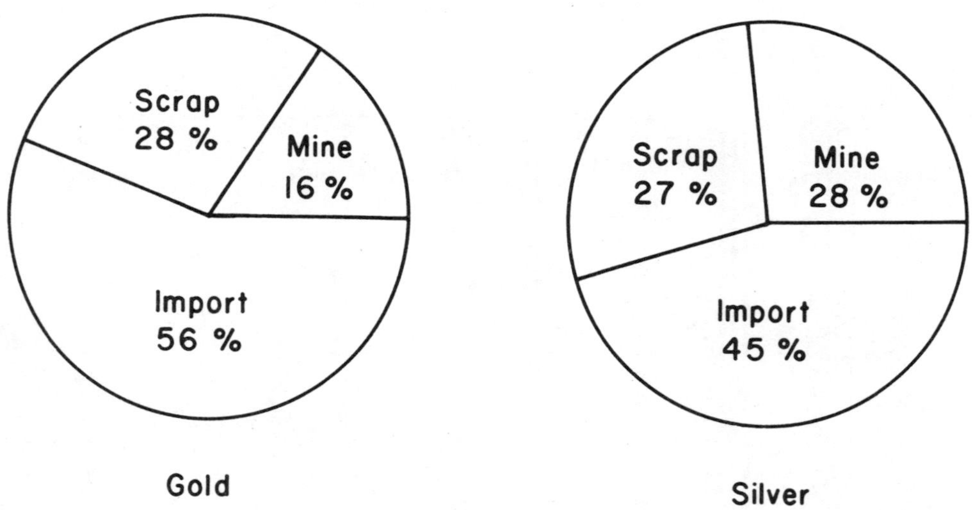

Figure 5. Gold and silver supply sources.

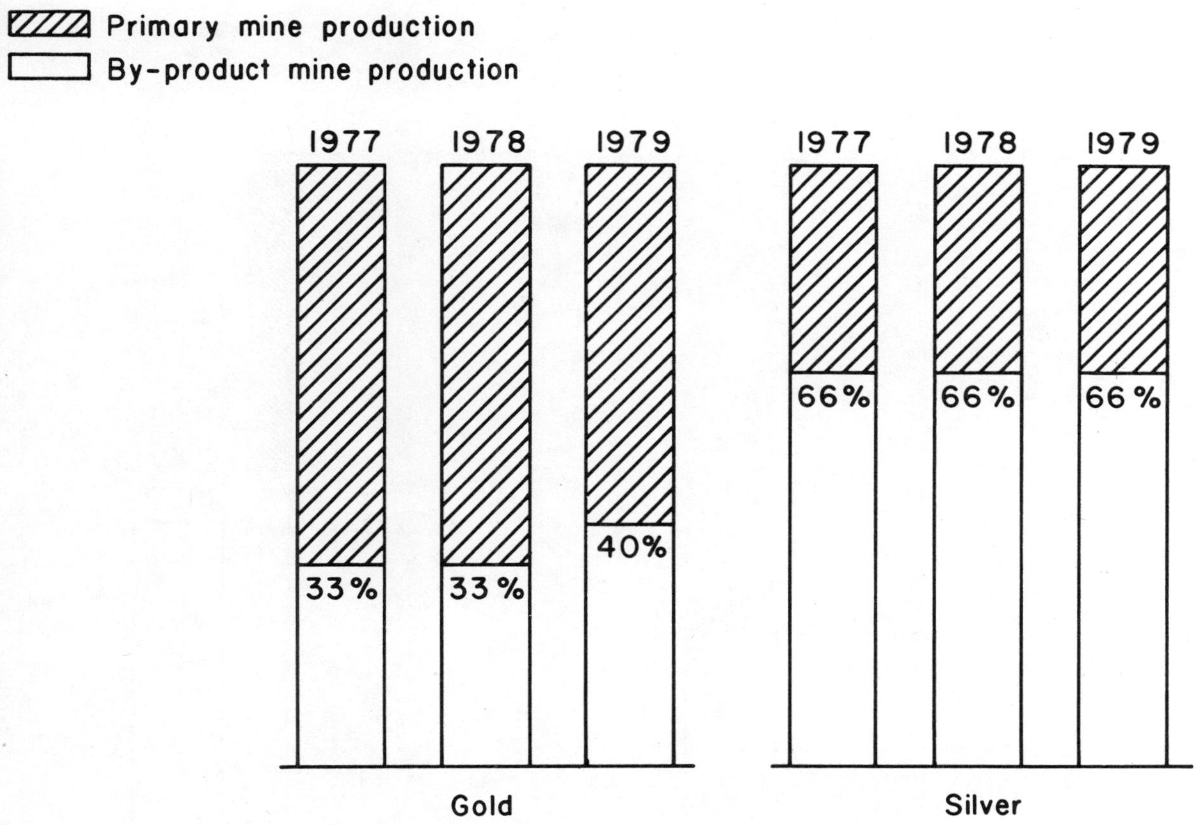

Figure 6. Gold and silver primary and byproduct production.

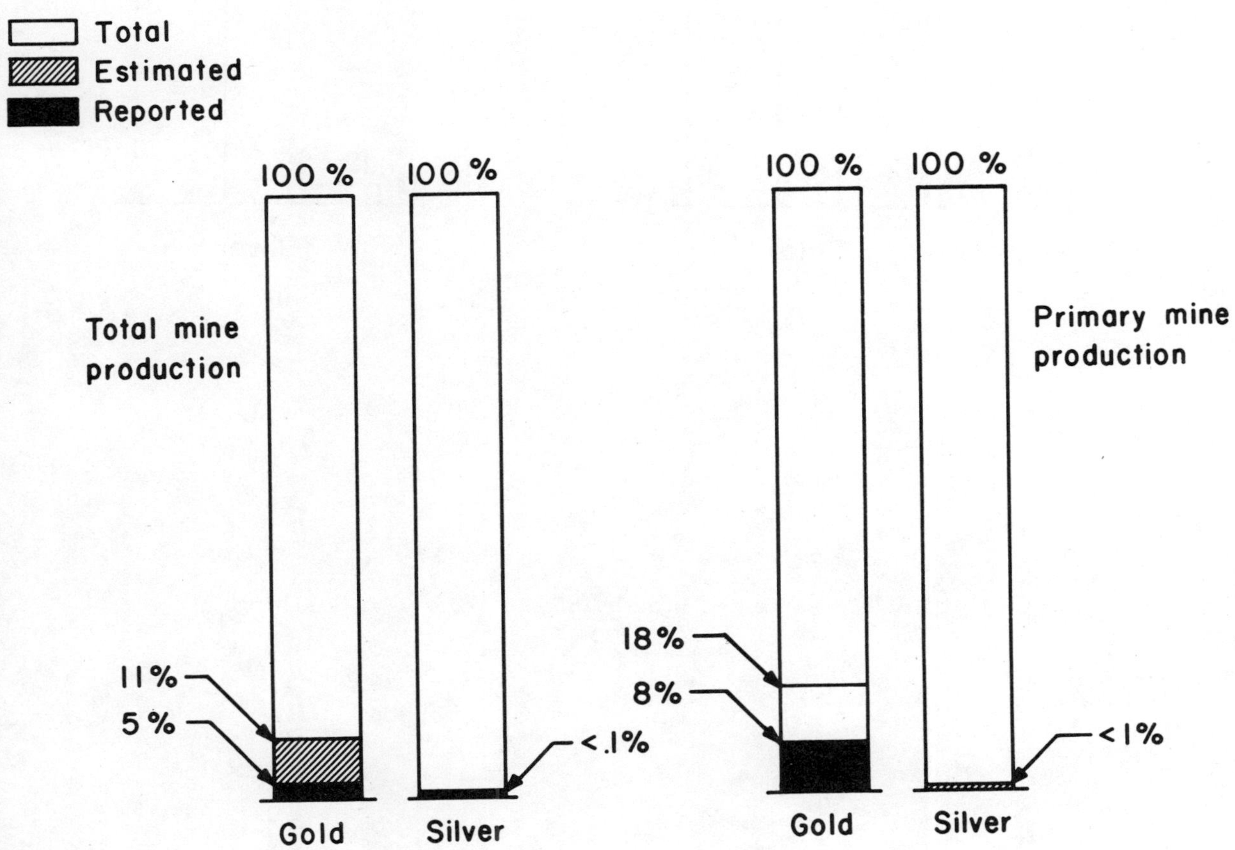

Figure 7. Gold and silver leaching production.

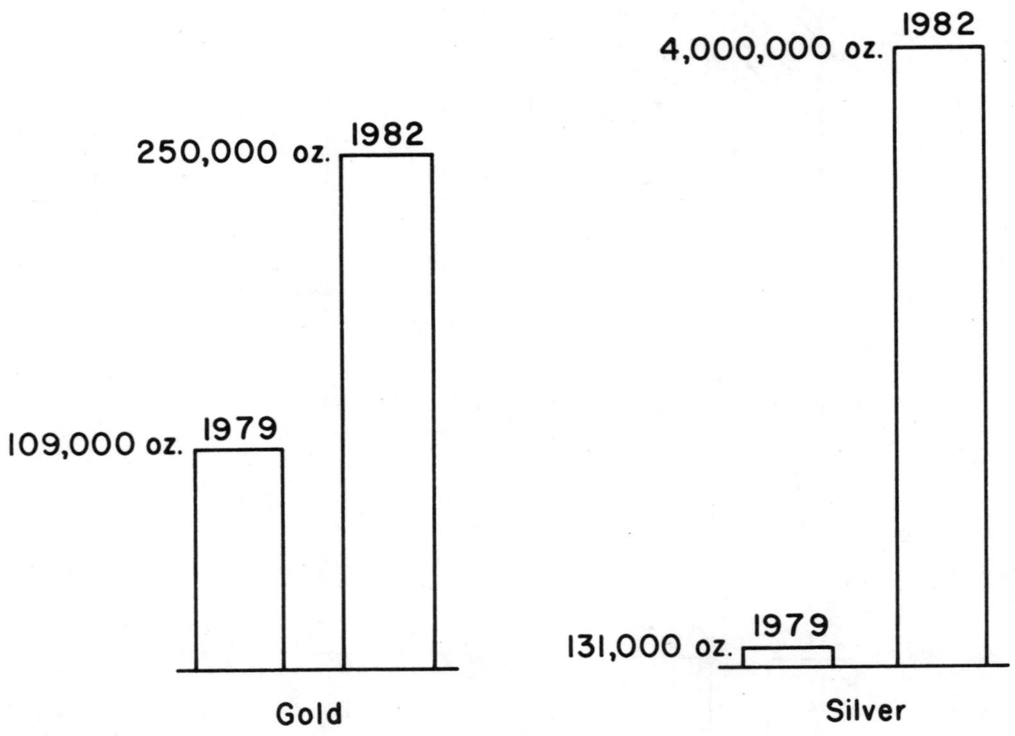

Figure 8. Estimated growth in gold and silver leaching.

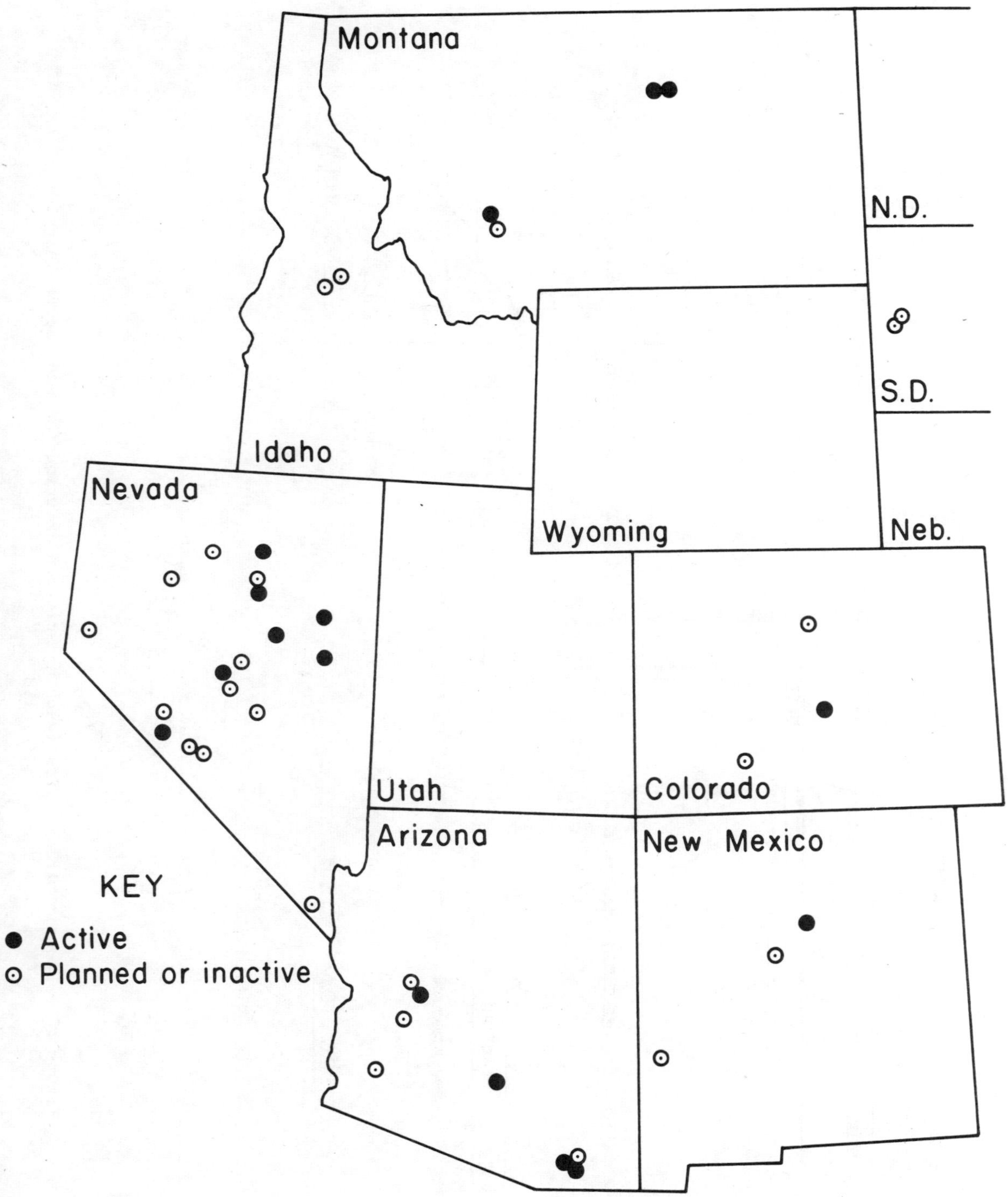

Figure 9. Gold and silver leaching operation location (Chamberlain and Pojar, 1981).

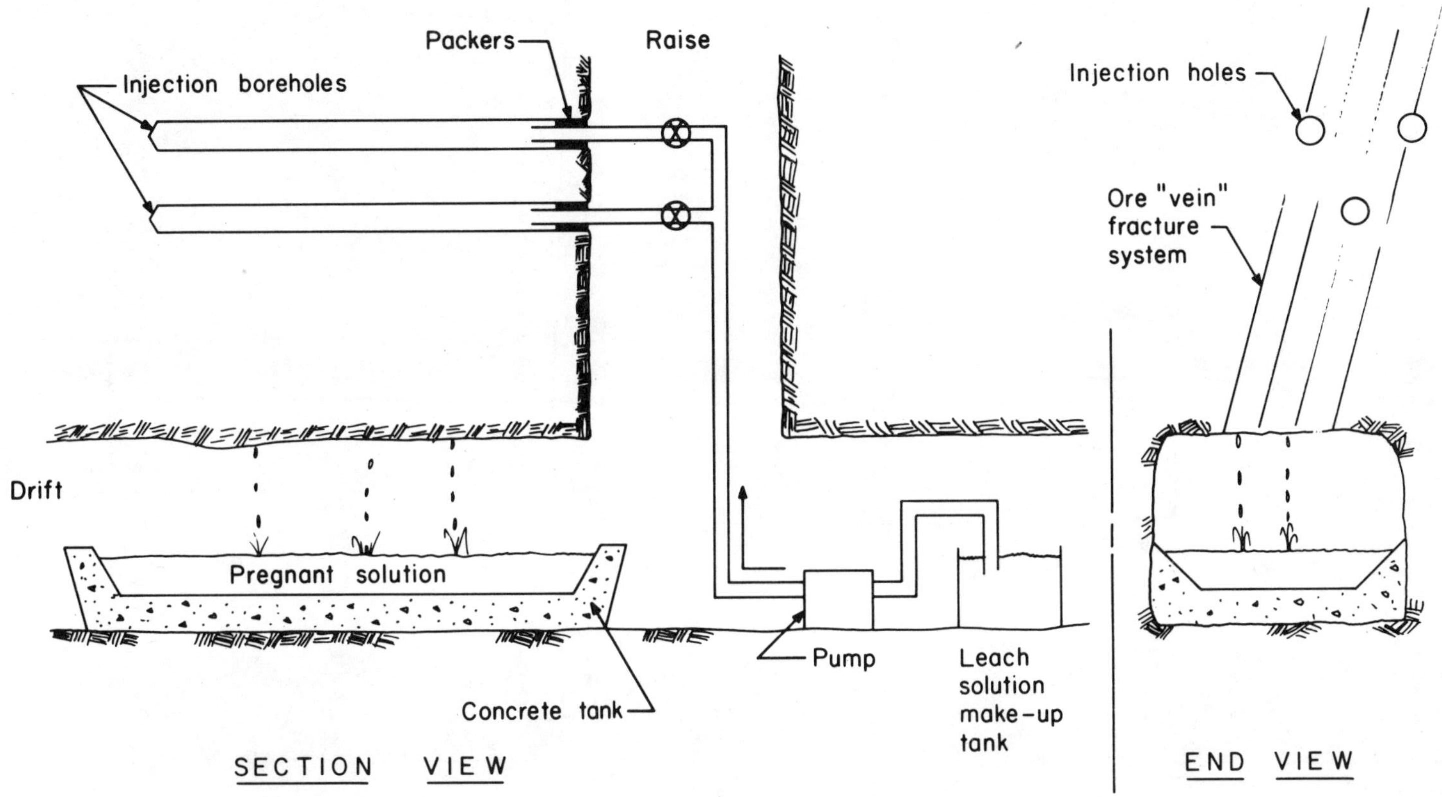

Figure 10. Experimental in situ leaching solution distribution and recovery system (Chamberlain and Pojar, 1981).

Figure 11. Poor percolation into heap (Chamberlain and Pojar, 1981).

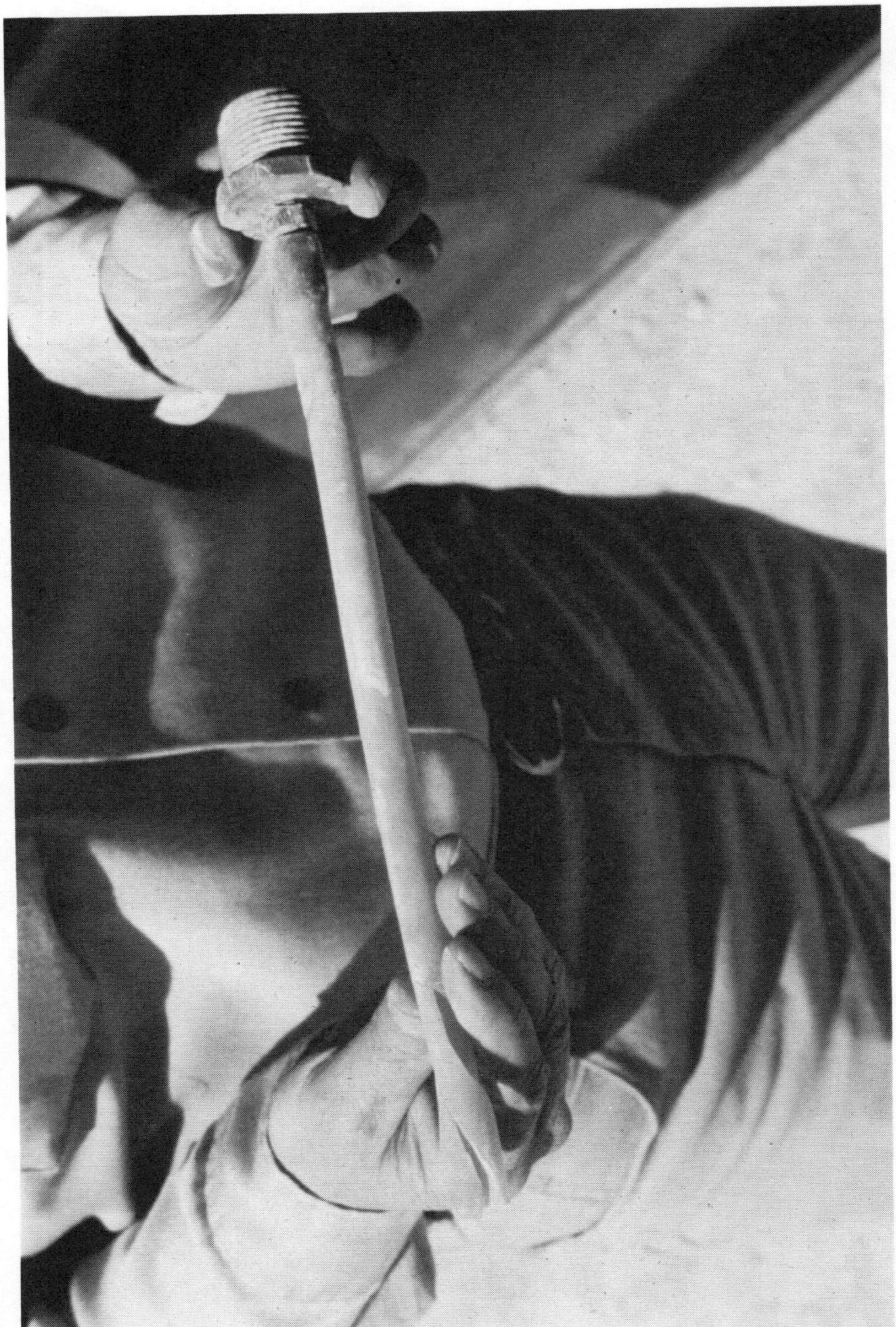

Figure 12. Bagdad Wiggler
(Chamberlain and Pojar, 1981).

Chapter 2

THE TREATMENT OF REFRACTORY GOLD ORES CONTAINING CARBONACEOUS MATERIAL AND SULFIDES

W. J. Guay
Newmont Exploration Limited
Tucson, Arizona 85704

ABSTRACT

Carbonaceous gold ores containing gold dispersed in sulfides are refractory to cyanidation because: (1) the carbonaceous materials adsorb gold during cyanidation; (2) a portion of the gold is so finely dispersed in the sulfide mineral lattice as to preclude contact with cyanide solution; and (3) a portion of the gold is chemically combined with carbonaceous material.

Techniques which have been used to treat these ores are: (1) roasting or chemical oxidation to decompose sulfide and carbonaceous materials; (2) flotation to reject carbon; (3) the use of kerosene or other reagents to passivate the adsorptive carbon compounds; (4) carbon or resin in pulp; (5) flotation of a bulk concentrate which is treated by one of the above methods; and (6) the use of chelating agents which have a stronger affinity for gold than the activated carbon in the ore.

Introduction

The adverse effects of carbonaceous material in the cyanidation of gold ores have been known for many years in many of the major gold producing camps according to R. W. Nice (1971), including the southwestern states, the California mother lode, the African gold coast, Russia, and the McIntyre Porcupine and Kerr Addison mines in Canada. Its effects have also been noted in the California mother lode by E. S. Leaver and J. A. Woolf (1930), and in the sedimentary Carlin type deposits of Nevada by the Nevada Mining Association (1980), W. J. Guay (1973 and 1980), B. J. Scheiner, et al (1971 and 1972), and A. S. Radtke and B. J. Scheiner (1970). These include the Carlin, Maggie Creek, Gold Acres, and Jerritt Canyon deposits.

In the McIntyre Porcupine Mine, R. W. Nice (1971), and at Gold Acres, mining was suspended altogether in areas containing too much carbonaceous ore. In some areas of the Carlin Mine, the carbonaceous ore is so deleterious to cyanidation that a mixture containing five per cent carbonaceous ore and 95 per cent standard cyanidable ore gives an extraction of 76.9 per cent gold vs. 93 per cent for 100 per cent standard ore.

At the McIntyre Porcupine Mine, R. W. Nice (1971), an "Activity Test" was run and at Carlin a "Preg Robbing" test is run to determine whether the ore is amenable to cyanidation treatment. The tests measure the adsorption of gold from pregnant solutions by ground samples over specified time periods. If the tests show any adsorption of gold the ore is not cyanided directly, but is separated for special pretreatments before cyanidation. Very rigid control of mining must be practiced to prevent any carbonaceous material from entering the cyanide circuit without pretreatment.

Mineralogy and Occurrence of Gold

The term "carbonaceous" has been loosely applied to ore constituents of widely varying characteristics. Laboratory investigations by A. S. Radtke and B. J. Scheiner (1970) of the U. S. Geological Survey and the U. S. Bureau of Mines on carbonaceous ores from Carlin showed that they contained:

(1) an activated carbon component capable of adsorbing gold chloride or gold chloride complexes from solution; (2) a mixture of high molecular weight hydrocarbons usually associated with the activated carbon components; and (3) an organic acid, similar to "humic acid" containing functional groups capable of interacting with gold complexes to form organic gold compounds. The structure of these compounds is not known. A possibility is that they are formed by chelation, where ligands such as N, S, or O in organic acids form stable gold chelates.

Activated carbon in the Carlin ores is defined by the U. S. Bureau of Mines as fine grained amorphous carbon which will adsorb stable gold complexes. Little is known of the structure of the hydrocarbons. Approximately 0.012% hydrocarbon was removed by organic solvents from one Carlin sample. Removal of the hydrocarbon resulted in less gold extraction when the ore was treated by cyanidation, suggesting that the hydrocarbon was adsorbed on the activated carbon. Organic acid in the carbonaceous ore was extracted by heating the ore in 1 N NaOH solution. The pulp was filtered and the filtrate acidified with HCl. After acidification a gelatinous precipitate formed, consisting of SiO_2, aluminum oxide, and 0.13-0.15% carbon. An infrared spectrum of the precipitate matches closely the spectrum of humic acid extracted from leonardite, a partially oxidized form of lignite from North Dakota. When the carbonaceous ore was treated with sodium hydroxide prior to cyanidation, the gold extraction increased significantly. When five grams of humic acid taken from leonardite was added to 100 grams of non carbonaceous cyanidable ore prior to cyanidation, gold extraction was reduced from 95 to 21 per cent. In another experiment the addition of five grams of humic acid to 25 ml. of a pregnant solution containing 0.032 grams of gold resulted in the humic acid adsorbing 84 per cent of the gold. All of these phenomena suggested that humic acid was chemically combined to the gold in the carbonaceous ores. Very strong supportive evidence that the gold was chemically combined was obtained in an experiment where a pulp of ground ore and ten per cent sodium chloride solution was electrolyzed for thirty minutes at pH 10.0 resulting in eighty per cent dissolution of the gold in the ore. With non carbonaceous ore containing gold in the metallic state, a similar experiment showed only eight per cent dissolution of gold.

Most of the Carlin carbonaceous ores contain from 0.4 to about 1.0 per cent organic carbon, with occasional samples as high as 4.0 per cent. Particle sizes vary from less than one μm to over 100 μm.

At the McIntyre Porcupine Mine, R. W. Nice (1971), the carbonaceous matter was identified as a disordered graphite structure with a small crystal size in the range of 0.002 to 0.005 μm.

The carbonaceous matter in the gold bearing carbon seams in the Witwatersrand ores of South Africa does not interfere with cyanidation, King (1949). It is described, J. E. Zumberge and A. C. Sigleo (1978), as a complex, solid, and solvent insoluble, polymer-like substance, containing mainly hydrocarbons and some organic sulfur and oxygen compounds.

Because the term "carbonaceous ore" cannot be rigorously defined, the classification of the ores at Carlin is arbitrary. Classification is by fire assay, atomic absorption (AA) assay on raw ore and on ore roasted at 500C, and by the "Preg Robbing" test previously described.
A positive "Preg Robbing" test is a rapid, certain way of identifying carbonaceous ores. However, some ores with a low carbonaceous content may not indicate carbon by the "Preg Robbing" test. The AA assay on raw or roasted samples is a hot cyanide leach of a small sample of pulverized ore, followed by an AA assay of the pregnant solution. If the AA assay on raw ore gives an extraction of over eighty per cent of the gold indicated by fire assay, the ore is amenable to cyanidation. Anything yielding less than eighty per cent extraction is refractory ore. If a refractory ore gives over eighty per cent extraction by the AA assay on a roasted sample, the refractoriness is due to either carbonaceous material or sulfides. If the AA assay on the roasted sample shows less than eighty per cent

extraction of gold, the refractoriness is due to gold lockup in the siliceous gangue. Work by A. S. Radtke and B. J. Scheiner (1970) of the U. S. Geological Survey and the U. S. Bureau of Mines classified Nevada ores containing 0.25-0.8 per cent organic carbon as carbonaceous ore and those containing 0.06 to 0.25 per cent organic carbon as non carbonaceous.

In addition to carbonaceous material, the Carlin type Nevada ores contain pyrite up to three per cent. Two types of auriferous pyrite at Carlin have been examined by a scanning electron microscope (SEM). Spheroidal pyrite occurs as isolated spherules or as clusters, showing high internal porosity, making it amenable to low temperature oxidation. It consists of numerous discrete pyrite crystallites, ranging from about 0.5 to nearly 2 μm in size that have aggregated to form spheroidal clusters that may range up to 20-30 um in aggregate dimension. Coarse, euhedral pyrite, 10 to 50 μm in size, is cubic with no internal porosity, thus making it much more difficult to oxidize at low temperatures. Particulate forms of gold in pyrite were not detected in the SEM studies, and the gold is therefore either in solid solution or as finely occluded particles less than 0.02 μm (the resolution limit of the instrument).

In microprobe studies by J. D. Wells and T. E. Mullens (1973), samples of ore from Cortez and Carlin showed major concentrations of 900 and more grams of gold per tonne occluded in spheroidal pyrite grains, and in smaller concentrations in thin rims around cubic grains.

Occurrences of auriferous pyrite are quite common and have been described by many investigators including P. Joralemon (1951), M. H. Haycock (1937), and F. L. Stillwell and A. B. Edwards (1946). At the Getchell Mine in Nevada pyrite concentrates commonly ran about 15 grams of gold per tonne.

As if the presence of carbonaceous material and auriferous pyrite was not enough of a problem, the Carlin ore also has from five to eight per cent illite, 1-9.5 per cent kaolin, and in some parts of the orebody a small amount of montmorillonite. These clay minerals can and have adsorbed the gold cyanide complex. A typical mineralogic analysis of a Carlin ore is:

	Per Cent
Dolomite	39
Calcite	14
Quartz	36
Illite	8
Kaolin	1
Pyrite	1
Carbon	1
	100

Treatment Methods

Flotation

The simplest method of treating a carbonaceous ore is the method described by R. W. Nice (1971) at the McIntyre Porcupine Mine where the carbon, which contains very little gold, is floated off and discarded. The remaining ore is then responsive to cyanidation. This does not work at Carlin because the carbonaceous compounds have a high gold content. A considerable amount of flotation work was done on Carlin ores, however, to produce a high grade concentrate for possible shipment to a smelter and a tailing which could either be discarded or directly cyanided. All of the concentrates contained both carbonaceous material and pyrite and showed low recoveries of gold and poor selectivity. This was because of insufficient liberation of carbonaceous material and sulfide from the gangue, even at 100 per cent minus 325 mesh.

In the treatment of an auriferous sulfide associated with carbonaceous shale from South Africa by American Cyanamid Company (1939), up to 77 per cent of the carbon was eliminated by depressing it with Cyanamid reagent 637 while floating the sulfides for 90.5 per cent gold recovery, and 20.4 to 1 ratio of concentration. Fortunately, the carbon in the Witwatersrand ores is in a form which does not interfere with normal cyanidation.

Blanking Agents

Blanking agents are used to pasivate the surfaces of activated carbon in the ore by being adsorbed preferentially to the gold cyanide complex. This method has been used by Kerr Addison Mines, R. W. Nice (1971). Another passivator, RV-2 (para nitro benzol azo salicilic acid), is reputed to have been used in Russia while heavy machine oils have also been used by another firm according to R. W. Nice (1971). These methods are not applicable to Carlin type ores because a considerable amount of the refractory gold is: (1) occluded in the pyrite lattice; and (2) combined with humic acid.

Cyanide Leaching with Carbon or Resin in Pulp

The addition of activated carbon to adsorb gold preferentially to the acivated carbon component in the ore was proposed by T. G. Chapman (1939 and 1950) who recovered the activated carbon by flotation. A modification of this procedure, using granular activated carbon instead of finely divided carbon, was tried on mildly refractory Carlin ores with partial success. Extractions of about seventy per cent of the gold were obtained by adding granular carbon to a a standard cyanide leach, compared to fifteen per cent extraction without the addition of carbon. For purposes of this presentation mildly refractory ores are defined as those requiring only ten to twenty Kg of chlorine per tonne in a cyanidation pretreatment to be described later, in order to yield 83 or more per cent gold in cyanidation. Similar results were obtained with Rohm and Haas anion exchange resin IRA 425.

These methods have the advantage of low capital and operating costs, and for ores whose only deleterious constituent is carbonaceous material in concentrations of less than 0.5 per cent organic carbon, may be the best way to go. For the Carlin ores, however, these methods were abandoned in favor of chemical oxidation which had the potential for much higher gold extractions.

Since the U. S. Bureau of Mines had shown that humic acid could be extracted by a hot solution of sodium hydroxide and since the use of a hot solution containing one per cent NaOH and 0.2 per cent NaCN was in commercial use to strip the gold cyanide complex from activated carbon at Homestake, K. B. Hall (1974), Newmont looked into using a solution of that composition to leach gold from highly refractory Carlin ores. These ores contained from 0.8 to 1.0 per cent organic carbon and from 1.5 to 2.5 per cent pyrite. Highly refractory ores are defined as those requiring over 100 Kg of chlorine per tonne in a pretreatment process in order to yield 83 or more per cent gold in cyanidation. The hot caustic leach used carbon in pulp, the theory being the same as with the previous carbon in pulp work with milder ores at ordinary temperatures. The gold would be desorbed from the natural activated carbon in the ore and transferred to the more strongly adsorbent added activated carbon until the concentration of gold on the added carbon was in chemical equilibrium with the concentration of gold on the natural carbon.

Extensive research was done on the method. With ores containing up to 0.8 per cent organic carbon and 1.5 per cent pyrite, the method worked quite well giving 83.0 per cent extraction of gold and a carbon loading of 1000 grams of gold per tonne in locked batch tests from an ore containing ten grams of gold per tonne. However, on the most refractory ores which contained 1.0 per cent organic carbon and 2.0 to 2.5 per cent pyrite the method gave only 78.3 per cent extraction of gold and a loading of only 333 grams of gold per tonne from an ore containing 10.3 grams per tonne, so the method was finally abandoned.

Chelating Agents

In early exploratory research at the U. S. Bureau of Mines in the late 1960s, malononitrile, $C_3H_2N_2$, was found to be a much more effective reagent in leaching gold from mildly refractory ores than was sodium cyanide because of its more powerful chelating properties. However, the resulting solutions would not give up their gold by zinc dust precipitation, and the effort was terminated. Malononitrile is very toxic and has a high vapor pressure at ordinary temperatures, with a boiling point of 218C at one atmosphere pressure. Its handling would therefore require extreme care in order to ensure safe working conditions.

Roasting

With Carlin ores, roasting drives off carbonaceous materials as CO and CO_2, and decomposes pyrite to iron oxides from which the gold can be leached. Extensive research at Newmont Exploration has shown that roasting requires a fine feed of minus ten mesh, at least four hours retention, and precise temperature and atmosphere in the furnace. Below about 500C or with the use of a slightly reducing atmosphere, roasting is incomplete, resulting in a drastic reduction in gold extraction. At 550C and above the pyrite is converted to a refractory form of hematite from which the gold cannot be leached with cyanide. There is also the need for expensive drying of the feed to the furnace, and the need for rigid emission controls on the furnace gases. Extractions of 85 to 87 per cent of the gold were the best that could be obtained from ores containing thirteen grams of gold per tonne. For these reasons and because of the very high capital cost requirements, roasting was abandoned as a suitable alternative at Carlin.

Chemical Oxidation

In a research program which began in 1967, the Reno, Nevada station of the U. S. Bureau of Mines showed that the deleterious effects of the carbonaceous materials in Carlin type carbonaceous ores could be overcome by oxidizing agents in aqueous pulps, B. J. Scheiner (1971). After extensive bench scale work by the U. S. Bureau of Mines, Newmont Exploration Limited, and Carlin Gold Mining Company, the choice of oxidation agents for the Carlin ores was narrowed to either chlorine or sodium hypochlorite generated in situ by electrolysis of an ore pulp containing brine, W. J. Guay (1973). After investigating these two methods on a small batch pilot plant scale over a period of several months in 1969, the chlorination method was selected. Electrolysis in pulp featured high capital and low operating costs while the use of chlorine featured low capital and high operating costs, chiefly for chlorine. The limited tonnage of ore available at that time dictated the use of chlorine. Since its startup in 1971, the chlorine plant has consistently outperformed its original expectations, giving gold extractions in the 83 to 90 per cent range at tonnages of 455 tonnes per day or more. Freeport Gold, Nevada Mining Association (1980), has announced that it will use chlorine to treat carbonaceous ores at Jerritt Canyon in Nevada.

Initially, the plant treated ores which required about 15 Kg of chlorine per tonne of ore to oxidize the carbonaceous material and the pyrite. In mid 1975 additional chlorination capacity was installed to treat ores which required up to 40 Kg of chlorine per tonne. Estimates in 1975 showed over 1,000,000 tonnes of refractory ore reserves which required over 50 Kg of chlorine per tonne. These larger quantities of chlorine are quite expensive and thus led to a search for alternatives.

A process was developed, W. J. Guay (1980), in which air is dispersed in an aqueous slurry of ground ore at forty to fifty per cent solids at temperatures of 80 to 86°C, until a considerable portion of the pyrite is oxidized to iron oxides. Some decomposition of carbonaceous materials also occurs, but it is necessary to follow the air oxidation with chlorination in order to complete the oxidation of the carbonaceous materials and pyrite. The process of aeration followed by chlorination has been named "Double Oxidation". The commercial plant was started up in December 1977. Extractions of gold of over 85 per cent at acceptable economics and under practical operating conditions have been shown to be feasible.

The key to the successful use of air to oxidize pyrite at Carlin is the porosity and large surface area of the spheroidal pyrite. The cubic pyrite is not completely oxidized, but fortunately it has a lower gold content. Other Carlin type ores from Nevada that we have seen have only the cubic pyrite and are not amenable to air oxidation of the pyrite. We have also seen Carlin type ores from Nevada that contain arsenopyrite, which is definitely not amenable to air oxidation. Other Carlin type ores have shown locking of pyrite in quartz. Thus, many Carlin type Nevada ores which are similar in gross composition and in megascopic properties are quite different when viewed in the submicron size range.

Complex Flowsheets

In the retreatment of amalgamation tailings from the mother lode in California, J. V. N. Dorr and F. L. Bosqui (1950), a sand slime separation was followed by cyanide leaching of the sands, and the flotation of sulfides and free gold from the slimes. The carbonaceous material was depressed and remained in the slime tailings while the concentrate was roasted and cyanided. In another similar scheme a coal tar product was added to the slimes as a blanking agent. This allowed cyanidation to take place without the precipitation of the gold on carbon.

References

1. American Cyanamid Co., 1939, "Ore Dressing Notes", Bulletin No. 9, Jan.
2. Chapman, T. G., 1939, "A Cyanide Process Based on the Simultaneous Dissolution and Adsorption of Gold", Rocky Mountain Fund Volume on Milling Methods, AIME.
3. Crabtree, E. H., Jr., Winters, V. W., and Chapman, T. G., 1950, "Developments in the Application of Activated Gold and Silver from Carbons", Trans. Mining Branch, AIME, Vol. 187.
4. Dorr, J. V. N. and Bosqui, F. L., 1950, "Cyanidation of Gold and Silver Ores", McGraw Hill, Second ed., p. 33.
5. Guay, W. J., and Peterson, D. G., 1973, "Recovery of Gold from Carbonaceous Ores at Carlin, Nevada", Trans. Soc. Min. Eng., AIME, Vol. 254, March, pp. 102-104.
6. Guay, W. J., 1980, "How Carlin Treats Gold Ores by Double Oxidation", World Mining, March, pp. 47-49.
7. Hall, K. B., 1974, "Homestake Uses Carbon-in-Pulp to Recover Gold from Slimes, World Mining, Nov., pp. 44-49.
8. Haycock, M. H., 1937, "The Role of the Microscope in the Study of Gold Ores", Can. Inst. Min. Met., Vol. 40, pp. 405-414.
9. King, 1949, "Gold Metallurgy on the Witwatersrand", Trans. Chamber of Mines, South Africa.
10. Joralemon, P., 1951, "The Occurrence of Gold at the Getchell Mine, Nevada", Economic Geology, Vol. 46, No. 3, pp. 267-309.
11. Leaver, E. S., and Woolf, J. A., 1930, "Retreatment of Mother Lode (California) Carbonaceous Slime Tailings", Technical Paper 481, U. S. Bureau of Mines, 20 pp.
12. Nevada Mining Assoc., 1980, "Freeport Reserves Expanded; Milling Upped to 2,750 Tons per Day", Bull., April.
13. Nice, R. W., 1971, "Recovery of Gold from Active Carbonaceous Ores at McIntyre", Can. Min. Jnl., June, p. 42.
14. Radtke, A. S., and Scheiner, B. J., 1970, "Studies of Hydrothermal Gold Deposition (1) Carlin Deposit, Nevada: The Role of Carbonaceous Materials in Gold Deposition", Economic Geology, Vol. 65, No. 2, March-April, pp. 87-102.
15. Scheiner, B. J., Lindstrom, R. E., and Henrie, T. A., 1971, "Processing Refractory Carbonaceous Ores for Gold Recovery", Journal of Metals, Vol. 23, March, pp. 37-40.
16. Scheiner, B. J., et al, 1972, "Extraction of Gold from Carbonaceous Ores: Pilot Plant Studies", Report of Investigation 7597, U. S. Bureau of Mines, February, 20 pp., 16 Figs.
17. Stillwell, F. L., and Edwards, A. B., 1946, "An Occurrence of Submicroscopic Gold in the Dolphin East Load, Fiji", Proc. Aus. No. 141, pp. 21-46.
18. Wells, J. D. and Mullens, T. E., 1973, "Gold-Bearing Arsenian Pyrite Determined by Microprobe Analysis, Cortez and Carlin Gold Mines, Nevada", Economic Geology, Vol. 68, pp. 187-201.
19. Zumberge, J. E., and Sigleo, A. C., 1978, "Molecular and Elemental Analysis of the Carbonaceous Matter in the Gold and Uranium Bearing Vaal Reef Carbon Seams, Witwatersrand Sequence", Minerals Sci. Engng, Vol. 10, No. 4, Oct., pp. 223-245.

Chapter 3

TREATMENT OF MANGANIFEROUS SILVER ORES FOR RECOVERY OF SILVER IN VIEW OF CHANGED PRECIOUS METAL ECONOMICS

Clement K. Chase
Mountain States Research & Development
Tucson, Arizona 85731

ABSTRACT

Since the dawn of cyanidation leaching techniques for precious metals, certain manganese-bearing silver ores have been a problem due to low silver extraction. Since simultaneous occurrence of silver and manganese is common in nature, this situation has made many otherwise attractive silver deposits uneconomic in the past.

Several older techniques have been useful but at $1.29 per ounce of silver, the economics were unfavorable. With the new look in silver prices and some improvements in processing technology, neglected silver districts offer new promise.

INTRODUCTION

Probably the first metal known to man was gold for reasons everyone will appreciate: the attractive color, the luster, malleability, durability, and its occurrence in the metallic state in nature. Gold has been cherished by man from antiquity. Silver and copper were next, since they occasionally occurred in native form. Later their reduction from oxidized ores was discovered, accidentally perhaps.

Processes for recovery of silver have an ancient lineage, dating back into antiquity. Amalgamation, retorting, chloride roasting followed by amalgamation, hyposulfite leaching, and the Patio Process were used from the 15th century until the beginning of the 20th century when an important revolution occurred: the almost universal advent of the cyanidation process.

Not all ores, however, were amenable to these earlier processes or even to the cyanide process. A common denominator for many of these refractory ores is the presence of manganese associated with the silver. The non-amenability of manganiferous silver ores to common cyanide extraction techniques is well known, although the exact mineralogical reason for this is elusive. However, it is known that dissolution of the manganese frees the silver for treatment and recovery by usual methods; salt roasting also does this. This paper attempts a survey of this potentially rewarding subject.

Another cause of poor recovery is silica encapsulation of the silver. This requires an extremely fine grind for silver liberation. Such fine grinding has been uneconomic in the past.

HISTORICAL PERSPECTIVE

In the past, many otherwise interesting silver deposits have lain untouched in the ground due to processing difficulties and resulting low extractions. Although several types of silver deposits are in this category, one of the most common (and most refractory) is the type where the silver is mineralogically tied up in compounds with manganese. These deposits frequently occur both in oxidized and sulfide form. This paper will deal with schemes for leaching the oxidized deposits since the sulfide deposits are often amenable to flotation concentration with subsequent smelting for product recovery. A discussion of the types of ores suitable for sulfide flotation is included.

The economics of special ore treatment methods were limited for many years by a silver price fixed by governmental fiat at $1.293 per troy ounce ($0.038 U.S. per gram). Since the advent of escalating oil prices due to the short-sighted actions of the OPEC cartel, with its resulting world depression, and the flight of investment money to the precious metals, the price of silver has risen to heights that make attractive more complex processing methods, even in the face of general world cost inflation. This justifies a new look at old methods. Ores with a value of $100 per ton are worth considerable processing effort.

SOME GENERAL TYPES OF SILVER ORES WITH SOME POSSIBLE EXTRACTION METHODS

1. Silver sulfide ores (amenable to flotation-smelting recovery techniques).
2. Oxidized, clean silver ores (amenable to cyanidation recovery).
3. Oxidized silver ores resistant to cyanidation recovery. Frequently these ores contain manganese.
4. Complex silver sulfide ores such as tennantite, tetrahedrite, and other combinations of silver with arsenic and antimony. (Suitable for flotation).
5. Argentiferous galena ores. (Flotation).
6. Argento-jarosite ores. (Difficult at best).
7. Silver ores with sulfide manganese. (Suitable for flotation).

POSSIBLE EXTRACTION METHODS

There are several approaches to the problem of silver recovery from manganiferous ores. These are listed with the knowledge that particular ores may be economically amenable to some unmentioned processes and that not all attempted processes are known to the author.

1. SO_2 leach, followed by thorough washing, liming to a suitable pH, and standard cyanidation. [18]
2. Chloride oxidation roasting followed by cyanidation or amalgamation. [68]
3. The venerable Patio Process - a combination of treatment with copper sulfate plus sodium chloride, followed by amalgamation for recovery of the silver.
4. The Washoe pan amalgamation process. (All processes using mercury today are environmentally difficult).
5. The Kerley thiosulfate leach process. This process is said to be effective for problem ores containing high percentages of manganese, copper, arsenic, tellurium, etc., or combinations of the above. The process is being offered by ThioTech of Sahuarita, Arizona. [51]
6. Direct smelter recovery of high grade ore or concentrates.
7. A chloridizing reduction roast (Segregation Process) followed by flotation. This could be particularly useful in a copper-silver-manganese ore such as that from the Berenguela Mine of the Lampa Mining Co.

near Puno, Peru. It could also be used on a combination of a refractory Ag ore mixed with an oxide copper ore. [78]

8. Flotation where suitable (sulfides).

9. Brine leach (as the tetrachloro-complex).

OPERATING PILOT AND PRODUCTION PLANTS

Two plants have operated in the past employing variations of roasting to render silver recoverable from manganiferous ores.

Salt Roasting Process at Kildún

At Cia. Minera Kildún y Anexas at Matehuala, S.L.P., Mexico, American Smelting and Refining Company (now ASARCO, Inc.) operated a commercial 400 MTPD salt roasting- cyanidation plant for eight years on ore averaging 2 grams gold per metric ton (0.06 oz. gold per short ton) and 500 grams silver per metric ton (14 oz. silver per short ton) in a gangue assaying 4 to 5% manganese. The New York silver price during this operating period was under 45¢ U.S. per ounce. The results of this operation were reported by R.J. Mellen in 1963. The thoughtful article by Mellen discusses problems frankly: corrosion due to high chloride liquors and roaster vapors, clinkers, the effect of grind on roasting success, and the operation of the calcine cooler.

In initial testing, straight cyanidation of this ore gave only about 20% silver recovery, although gold recovery was 86%. Salt roasting tests indicated that gold recovery could be improved somewhat and silver recovery raised to 87%, a worthwhile accomplishment, particularly in the present international precious metal price situation:

Time Period	Au Value/MT Ore	Ag Value/MT Ore
1937-1944	$ 2.00	$ 6.43
1980	34.00	286.00

The costs of roasting do not appear as heavy as they did 40 years ago.

In the described operating plant, the ore was roasted with 5% NaCl in two direct-fired rotary kilns with retention time of 30 minutes at a temperature over 800°C. The calcines were quenched, ground in mill solution, the pregnant liquor was separated by CCD, and the values precipitated by zinc dust. Operational problems were many but were overcome by a determined staff.

Techniques developed in the late 1940's and early 1950's for the salt roasting of vanadium ores should be of ready applicability to salt roasting of silver ores. Use of cast stainless steel rabble arms in a skinner-type hearth roaster, continuous quench of calcine directly from the roaster, and various corrosion-proofing techniques come to mind. Use of modern techniques, machinery, and plastics, coupled with current precious metal prices, should make an ore deposit such as the one at Matehuala very attractive today.

The Segregation Process

The Segregation Process was originally a method for treatment of oxide copper ores but was proved by tests to be suitable for silver-copper ores, also. The process roasts an ore at 700-800°C with, usually, sodium chloride and a reducing agent such as coke, charcoal, or recently, petroleum coke. The copper (and silver, if present) are volatilized, migrate, and are reduced on the carbon particles as metallic copper/silver alloy, whence they can be floated or leached. The flotation concentrate is commonly sent to a smelter as a high-grade concentrate.

The process was originally acidentally discovered in 1923 in experimental work on the oxide copper ores from the Sagasca deposit in Chile. These ores contain a natural level of sodium chloride and it was found that in attempting to reduce the copper with coal, the copper was reduced and then migrated to the coal surfaces where it was deposited as metallic copper.

The process was further developed in the period between 1923 and 1931. A series of patents were secured by the original Minerals Separation Company. Commercial development of the process was considered in 1931 in several locations but gathering world depression was not propitious for such efforts. However, several large scale pilot plants were operated. The following is a very brief history of pilot operations.

Segregation Process at Berenguela Mine

The second plant to be operated successfully using a roasting process was in reality conceived as a pilot plant but it was a large one and was skillfully operated by a knowledgeable staff. It used the Segregation Process for an unusual copper-silver manganiferous ore near Puno, Peru from the Berenguela Mine which was once operated by the Lampa Mining Co. (British).

Mineral Separation Company in Southern Rhodesia, Africa

A 50 ton per day pilot plant was operated in Southern Rhodesia by the Minerals Separation Company using a hearth roaster in a two-stage process. The plant operated for four months in 1931, treating a 5% oxide copper ore with a recovery of 87% by flotation in a concentrate assaying 68% copper.

Union Minière de Haut Katanga Present-Day Zäire in Africa

This plant had a capacity of 350 TPD. It operated for several months in 1931 using two direct-fired rotary kilns in a two-stage process.

Akjoujt Mine in Mauritania, West Africa

A small pilot plant of 50-75 TPD capacity was operated in Mauritania in 1959 on a problem oxide copper ore at 3% copper and 3 grams of gold per ton. The recovered concentrate assayed 60-70% copper and 2 ounces gold per ton. A commercial plant was planned and built, but failed because of operating problems.

Lake Shore Mill Transarizona Resources Corp., Arizona

A commercial operation of the Segregation Process was attempted south of Casa Grande, Arizona for several years beginning in 1960. Original operation was in an indirectly heated rotary kiln. Later, the plant was converted to a direct-fired operation. Recovery was near 85% with a concentrate running 50% copper. Operations were not sufficiently economic for a continued run, partly due to dust losses.

N'Changa, Anglo-American Corporation Southern Rhodesia

The process used here was named TORCO - Treatment of Refractory Copper Ores. It used a fluid solids roaster. A 10 TPD pilot plant was successful but a larger plant at 500-1000 TPD suffered from operational difficulties and did not run for long. A two-stage process was used; however, problems developed in transferring hot material from one stage to the next.

This experience with various copper operations using the Segregation Process is cited for reference value in small and large operations. Considerable operational experience has been gathered over the years on this intriguing metallurgical process.

It now appears that there are four effective methods for recovering silver from manganiferous ores: the sulfur dioxide leach, the oxidizing salt roast, the segregation process, and Thio Tech's thiosulfate leach. The segregation process has been demonstrated as being suitable for manganiferous silver ores by R.D. Groves and by operations on the Berenguela ore.

Fresnillo, Mexico SO_2 Leaching Process

This was a commercial application of the McClusky process using the sulfur dioxide preparatory leach for treatment of manganiferous silver ores. It was used on a portion of the silver ores treated and the SO_2-leached pulp was limed and joined with the other ground ores for cyanidation.

Candelaria - Modern Heap Leaching for Silver

A modern heap leach is in operation at the Candelaria Mine near Hawthorne, Nevada. The operating general partner is Occidental Minerals Corporation. Tonnage treated is 8000 per day. The ore is generally believed to be manganiferous. The mining operation is on such a scale that substantial silver production is assured.

APPLICABLE CHEMISTRY OF SILVER EXTRACTION

Chloride Roasting Followed by Cyanidation:

$$Ag_2S + 2NaCl + 2O_2 \longrightarrow 2AgCl + Na_2SO_4$$

$$AgCl + 2NaCN \longrightarrow NaAg(CN)_2 + NaCl$$

Chloridizing Leach:

$AgCl + 3\ NaCl \longrightarrow Na\ AgCl_4$ (soluble in excess brine)

Chemistry of Manganese Removal From Tie-Up With Silver:

Also $MnO_2 + SO_2 \longrightarrow MnSO_4$

(somewhat pH dependent)

$$MnO_2 + 2SO_2 + 1/2O_2 + H_2O \longrightarrow MnSO_4 + H_2SO_4$$

(The generated H_2SO_4 helps to dissolve divalent Mn).

$$MnO + H_2SO_4 \longrightarrow MnSO_4 + H_2O$$

(also soluble in H_2SO_4 is the iron portion of jarosite).

$$Fe_2O_3 + 3H_2SO_4 \longrightarrow Fe_2(SO_4)_3 + 3H_2O$$

PRODUCT RECOVERY METHODS

In all the above methods where cyanidation is used, product recovery is frequently accomplished by zinc dust precipitation with subsequent fire refining. In cases of gold-bearing ores, concentration of values from low-grade cyanide solutions by adsorption on activated charcoal with subsequent electrowinning of a doré bullion is a useful method but in the case of silver ores running several or more ounces per ton, the sheer bulk of the necessary carbon points to the direct zinc dust precipitation method as preferable. An alternate precipitation method involves the use of NaHS or Na_2S as suggested by Reno USBM to precipitate a silver sulfide product. Use of this method has been held back by poor settling and filtration rates for the silver sulfide precipitate and the insidious toxicity of possible hydrogen sulfide gas, to say nothing of hydrogen cyanide.

The recovery methods employing sulfur dioxide to dissolve the manganese offer the possibility of a manganese recovery step. This may be economic in those ores containing substantial manganese values. Parenthetically, it is felt that even very low values of manganese may be sufficient to render the silver difficult to dissolve with cyanide if the manganese and silver are syngenetic.

BIBLIOGRAPHY

Considerable effort has been given to collection of an adequate bibliography since it is felt that this may be the major contribution.

The author wishes to acknowledge the fine cooperation of the following individuals in the task of getting these references together: Norman E. Dausinger, Tucson, AZ; Judith A. Eisele, Reno, NV; James B. Fletcher, Miami, AZ; Douglas N. Halbe, Denver, CO; Thomas Henderson, Jr., Tucson, AZ; William C. Larson, Twin Cities, MN; Edward Martinez, S. Plainfield, NJ; George M. Potter, Tucson, AZ; Barry N. Watson, Tucson, AZ, and Norman L. Weiss, Tucson, AZ.

The bibliography is quite complete in regard to the Segregation Process (reducing chloride roast) since these references are difficult to find in print. The process combines a roast for liberation of metal values with the efficient flotation process for recovery of the values and is of demonstrated effectiveness in recovery of silver from manganiferous ores.

Paul Eimon's paper has a rather complete and excellent geologically-oriented bibliography appended.

BIBLIOGRAPHY

[1] Adamson, R. J., "Gold Metallurgy in South Africa" Cape and Transvaal Printers Ltd., Cape Town, South Africa, 1972, 452 pp.

[2] Addison, R., "Gold and Silver Extraction from Sulfide Ores", The Mining Congress Journal, October 1980.

[3] Agey, W. W., J. V. Batty, H. W. Wilson, and W. J. Wilson, "Beneficiation of Calico District, California, Silver-Barite Ores", U.S. Bureau of Mines RI 7730, 1973.

[4] Anon., "Metallurgical Progress", Mining Magazine, 36, 1927, p. 131.

[5] Anon., "Segregation Process for Copper Ores", Mining Magazine, 42, Jan. 1930, p. 59.

[6] Anon., "Concentrating Copper Ores - A New Process", Mining Jnl. (London), 168, Jan. 11, 1930, p. 33.

[7] Anon., Editorial, Mining Magazine, 42, 1930, p. 331.

[8] Anon., "U.S.B.M. Charts Segregation Process", Eng. & Min. Jnl., 160, Nov. 1959, pp. 98-99,

[9] Anon., "Segregation Process to be Used at New Copper Project (Transarizona)", Mining World, 21, Dec., 1959, p. 19.

[10] Anon., "Arizona Copper Silicates Respond to Segregation", Eng. & Min. Jnl., 161, Nov. 1960, pp. 86-87.

[11] Anon., "Flowsheet on the Segregation Process", Deco Trefoil, Bull. No. M7-F94, Jan.-Feb., 1965, pp. 17-18.

[12] Anon., "Bureau of Mines Copper Recovery Process to Get Large-Scale Trial", News Release by U.S. Dept. of Interior, Feb. 20, 1965.

[13] Apell, G. A., S. W. Hazen and E. G. Howe, "Lake Valley Manganese Deposits, Sierra County, N. Mex.", U. S. Bureau of Mines RI 4099, 1947.

[14] Anderson, James A., Reading Notes on: "The Geochemistry of Manganese Oxides - Guide to Sulfide Ore Deposits", Harvard University, PhD Thesis, 1964.

[15] Back, A. E., S. F. Ravitz and K. E. Tame, "Formation of Dithionate and Sulfate in the Oxidation of Sulfur Dioxide by Manganese Dioxide and Air", U. S. Bureau of Mines RI 4931, 1952.

[16] Bechaud, L. J., Jr., "Minerals Beneficiation", Review, Min. Cong. Jnl., 47, Feb. 1961, p. 63.

[17] Belladen - Noli et Sommariva Gaz. Chim. - Ital. 58 (1928) 443-9.

[18] Bender, F. N. and Carl Rampacek, "Percolation Leaching of Manganese Ores With Sulfur Dioxide", U. S. Bureau of Mines RI 5323, 1957.

[19] Berezowsky, R.M.G.S. and V. B. Sefton, "Recovery of Gold and Silver from Oxidation Leach Residues by Ammoniacal Thiosulfate Leaching", AIME Annual Meeting, Feb. 18-22, 1979, New Orleans, La.

[20] Blessing, Lee R., "Experimental Work on Manganese Silver Ores", University of Arizona, MS Thesis, 1936.

[21] Bodart, J., Contribution a l'etude des equilibres heterogenes Hydrolyse des chlorures metalliques par la vapeur surchauffee. These Bruxelles, probablement 1928.

[22] Bouton, C. M., W. C. Riddell and L. H. Duschak, "Silver in Chloride Volatilization", TP 317, U. S. Bureau of Mines (1924).

[23] Bradford, R. H. and C. M. Mac Farlane, "A Consideration of Some of the Problems of the Chloride Volatilization Process", TP No. 3, Mining and Metallurgical Investigations, Utah Engineering Exp. Station (1928).

[24] Brewer, L., "The Fusion and Vaporisation Data of the Halides", In L.L. Quill, The Chemistry and Metallurgy of Miscellaneous Materials, Thermodynamics, McGraw Hill (1950).

[25] Brewer, L. and N. L. Lofgren, "The Thermodynamics of Gaseous Cuprous Chloride", Monomer and Trimer, J. Am. Chem. Soc. 72 1950) 3038.

[26] Carpenter, J. A., R. R. Elliott and B. F. W. Sawyer, "The History of Fifty Years of Mining at Tonopah, 1900-1950", Univ. of Nev. Bull., Vol 45, No. 3, 1953, 153 pp.

[27] Clevenger, G. H. and M. H. Caron, "The Treatment of Manganese-Silver Ores", U.S. Bureau of Mines Bulletin 226, 1925, 110 p.

[28] Coffin, L.D., Treatment of Roasted Chalcopyrite in the Anglo American Corporation Mini-Segregation Plant in the Recovery of Metallic Copper, AMAX Report, Project No. 912 (753), Report No. 74-2, June 4, 1974.

[29] Crawford, W.E., AIME Transactions #112, 1934. (McClusky Process, Operated at Fresnillo, Mexico).

[30] Diaz, C. M., "Mechanism for the Segregation Process and Its Potential Application to Nickel Ores", M.S. Thesis, Dept. of Mineral Engineering, Columbia Univ., 1958, pp. 1-59.

[31] Dorr, J.V.N., "Cyanidation and Concentration of Gold and Silver Ores", McGraw-Hill Book Co., New York, 1936, 1st Edit.

[32] Dorr, J.V.N. and F.L. Bosqui, Cyanidation and Concentration of Gold and Silver Ores, McGraw-Hill Book Co., New York, 1950, 511 pp.

[33] Egleston, T., The Metallurgy of Silver, Gold, and Mercury in the United States, Vol. I, Silver, John Wiley & Sons, New York, N.Y., 1887.

[34] Eimon, P.I., "Bulk Silver Deposits in Volcanics, North American Cordillera - The Evolving Geologic Model, Publications, Predictions".

[35] Fichte, R., Die Thermodynamischen Eigenschaffen du Metallchloride, VEB. Verlag Technik Berlin (1953).

[36] Freeman, G. A., C. Rampacek and L. G. Evans: "Copper Segregation at the Lake Shore Mine", Jnl. of Metals, 13, May, 1961, pp. 370-372, Mining Eng., 13, Oct., 1961, pp. 1153-1155.

[37] Giudice, Phil, "Silver Considerations on Exploration", (paper) November 4, 1977.

[38] Glassner, A., "The Thermochemical Properties of the Oxides, Fluorides and Chlorides, to 2500°K", U.S. Atomic Energy Commission, Report ANL.5750, The Supt. of Documents, Washington, U.S.A. (1959).

[39] Gonser, B. W. and E. E. Slowter, "The Coating of Metals With Tin From the Vapour Phase (Stannizing)". Research at Battelle Memorial Institute, Int. Tin Research Council, Series A No. 76 (1938).

[40] Groves, R. D., "The Segregation Process Applied to a Manganese-Silver Ore", U. S. Bureau of Mines Internal Memorandum, Personal communication.

[41] Guilissen, "Theorie du Procede de Segregation", Courte Note Pour l'Union Miniere du Haut Katanga, March 23 (1928).

[42] Hamilton, E.M., "Manual of Cyanidation", McGraw-Hill Book Company, New York, N.Y., 1920.

[43] Harder, Edmund Cecil, "Manganese Deposits of the United States", USGS Bulletin 427, 1910.

[44] Havens, R., S. J. Hussey, J. A. McAllister and K. C. Dean, "Beneficiation of Oxide Manganese and Manganese-Silver Ores from Southern Arizona", U. S. Bureau of Mines RI 2024, Feb. 1954.

[45] Heinen, H. J., D. G. Peterson and R. E. Lindstrom, "Silver Extraction from Marginal Resources", Oral Presentation, The 104th TMS-AIME Annual Meeting, New York, N.Y., Feb. 16-20, 1975.

[46] Hedley, N. and H. Tabachnick, "Chemistry of Cyanidation", American Cyanamid Bulletin 23, June 1958.

[47] Henley, K.J., "Gold-Ore Mineralogy and Its Relation to Metallurgical Treatment", Minerals Sci. Engng., Vol. 7, No. 4, 1975, pp. 289-312.

[48] Hewett, D.F., "Manganese is a Clue to Deep Base and Precious Metals", Mining World, July 1963.

[49] Hewett, D. F., "Silver in Veins of Hypogene Manganese Oxides", Geological Survey Circular 553, 1968.

[50] Iwasaki, I., J. Takahashi and H. Kahata, "Extraction of Nickel from Iron Laterites and Oxidized Nickel Ores by a Segregation Process", AIME, New York, March 1966.

[51] Kerley Chemical Company, Sahuarita, Arizona 85629, Personal communication.

[52] Kroll, W. J., "Chlorine Metallurgy", Metal Industry, Sept. 26, Oct. 3, 10, 17, 24, 31, Nov. 7 (1952).

[53] Kubaschewski, O. and E. Li Evans, "Metallurgical Thermochemistry", Pergamon, 3rd Ed. (1958).

[54] Kudryk, V., "Silver from Low-Grade Refractory Ore", AIME Annual Meeting, Feb. 15-19, 1970, Denver, Colorado.

[55] Layng, H. R., "Chloridizing Volatilization", Engineering and Mining Journal Press, Nov. 12, 1921, T. P. Holt, Engineering and Mining Journal 117 (1924) 922.

[56] Leaver, E. S. and J. A. Woolf, "Re-Treatment of Mother Lode (California) Carbonaceous Slime Tailings", Bu Mine TP 481, 1930, 20 pp.

[57] Letcher, O., "Minerals Separation Devices Process for Mixed Copper Ores," Eng. & Min. Jnl., 128, Nov. 16, 1929, p. 786.

[58] Liddell, D. M., Handbook of Nonferrous Metallurgy, McGraw-Hill Book Co., New York, N.Y. 1945, "Brine Leaching of Silver Ores", pp. 530-534.

[59] Liddell, D. M., Ibid, "Manganese-Silver Ores", pp. 535-536.

[60] Lloyd, P. J. D., "Maximizing the Recovery of Gold From Witwatersrand Ores", Minerals Sci. Engng., Vol. 10, No. 3, 1978, pp. 208-221.

[61] Mackay, K.E. and N. Gibson, "Development of the TORCO Plant at Rhokana Corporation, Limited, Zambia", Transactions of the Institute of Mining and Metallurgy, Volume 77, pp. C19-31, 1968.

[62] Marcuson, S.W., "Application of the Segregation Process to Roasted Copper Concentrates", Mineral Science Engineering, Vol. 12, No. 1, January 1980.

[63] Martinez, E., "The Copper Segregation Process - Review of the Literature", ASARCO internal memo, 1965.

[64] Martinez, E., "Thermoanalysis of Copper Segregation", Mining Engineering, August 1966, pp. 70-73.

[65] Martinez, E., "The Copper Segregation Process Studied by Thermoanalysis", Transactions SME/AIME, Vol. 238, June 1967, pp. 172-179.

[66] McKinney, W.A. and L.G. Evans, "Segregation of Copper Ores by Direct-Firing Methods", U.S. Bureau of Mines, RI 6215 (1962).

[67] McKinney, W.A. and P.T. Waddleton, "Use of Various Salts as Copper-Volatilizing Agents in the Segregation Process", U.S. Bureau of Mines, RI 6044 (1962).

[68] Mellen, R.J., "How Silver-Gold Ores Respond to Salt Roasting-Cyanidation", Engr. Mng. Jrnl., Vol. 164, No. 4, pp. 76, 77, 83, April 1963.

[69] Mellen, R. J., "Crushing Changes Necessitated by Roasting Conditions at Compañía Minera Kildún y Anexas", Mining Technology, March 1945.

[70] Mellen, R. J., "Compania Minera Kildún y Anexas - 300 Ton Salt Roasting and Cyanidation Mill", American Smelting & Refining Co. memo.

[71] Meunier, J., "Gazeification et Oxydation des Combustibles", Masson, Paris (1958).

[72] Moulden, J.C. and B. Taplin, Metals Production, Limited, British Patent 250,991, 1924.

[73] Nolan, T.B., "The Underground Geology of the Western Part of the Tonopah Mining District, Nevada", Univ. of Nev. Bull., Vol. 24, No. 4, 1930.

[74] Opie, W.R., L.D. Coffin, and D.C. Cusanelli, "A Minimum Pollution Low-Energy Pyrometallurgical Process for Treating Chalcopyrite Concentrates", The Extractive Metallurgy of Copper, American Institute of Mining, Metallurgical, and Petroleum Engineers, Volume 1, Chapter 20, pp. 416-426, 1976.

[75] Opie, W.R., "Pilot-Scale Test to Evaluate the TORCO Segregation Unit for Treating Chalcopyrite Calcine", AMAX Interoffice Memorandum to J. Towers, December 6, 1974.

[76] Opie, W.R., L.D. Coffin, D.L. Armant, and Otto F. Cimler, "Selective Recovery of Copper from Copper-Nickel Sulfide Concentrates by Applying the Segregation Technology", Metallurgical Transactions B, Volume 10B, pp. 27-32, March 1979.

[77] Pinkney, E.T. and N. Plint, "Treatment of Refractory Copper Ores by the Segregation Process", Transactions of the Institute of Mining and Metallurgy, Volume 76, pp. C114-132, 1967.

[78] Pollandt, F. and M. E. Pease, "Extraction of Copper and Silver by the Segregation Process in Peru", Bulletin of the Institution of Mining and Metallurgy No. 646, September 1960. Discussion - Bull. 649 (Dec. 1960) 652 (March 1961).

[79] Potter, G. M., and Roshan B. Bhappu, "The Chemistry of Gold and Silver Recovery from Natural Resources", Mountain States Research and Development paper presented at the American Chemical Society Annual Meeting, Anaheim, Calif., 1978.

[80] Potter, G. M., "Some Trends in Gold and Silver Metallurgy" - Original: "Algunas Tendencies en la Hidrometalurgia del Oro y Plata", Memoria de la Asociación de Ingenieros de Mina, Metalurgistas y Geologos de Mexico, Guanajuato, GTO, Mexico, Nov. 1978.

[81] Potter, G. M., "Merrill-Crowe Precipitation of Precious Metals by Zinc Dust", presented at Hydrometallurgy Division, Spring Meeting, Arizona Conference, SME-AIME, May 10, 1980.

[82] Ralston, O. C., "Segregation Process for Oxidized Copper and Other Ores", Unpublished review, 1957.

[83] Rampacek, Carl, H. C. Fuller and J. B. Clemmer, "Operation of a Dithionate-Process Pilot Plant for Leaching Manganese Ore From Maggie Canyon Deposit, Artillery Mountains Region, Mohave County, Arizona", U.S. Bureau of Mines RI 5508, 1959.

[84] Rampacek, C. and W. A. McKinney, "The Copper Segregation Process", Paper presented at AIME meeting in New York, Feb. 1960 (Preprint No. 60B61).

[85] Rampacek, C., W. A. McKinney and P. T. Waddleton, "Treating Oxidized and Mixed Oxide-Sulfide Copper Ores by the Segregation Process", U. S. Bureau of Mines, RI 5501 (1959).

[86] Rey, M., "Le Procede de Segregation des Minerais de Cuivre Oxydes Pauvres", Congres International des Mines, de la Metallurgie et de la Geologic Appliquee, Paris, October 1935, Egalement: Revue dé Metallurgie 33 (1936), 293.

[87] Rey, M., "The Segregation Process for Low-Grade Copper Oxide Ores", 7th Int. Cong. Min. Met. & Appl. Geol., Met. Ses., Vol. II-3, Paris, Oct. 20-26, 1935, pp. 55-62; Rev. Met., 33, 1936, pp. 295-302.

[88] Rey, M., "The Enrichment of the Copper Minerals of the Akjoujt Mine in Mauritania", Rev. Ind. Min., 41, Aug., 1959, pp. 667-669.

[89] Rey, M., "Mines de Cuivre de Mauritanie-Copper Segregation Process Testing", 1954-1962, (January 1967), Report for Anglo American.

[90] Rey, M.R.W., "Early Development of the Copper Segregation Process", Transactions of the Institute of Mining and Metallurgy, Volume 76, pp. C101-107, 1967.

[91] Rey, M., "Memoirs of Milling and Process Metallurgy #4 - Segregation Processing", IMM Transactions C, Vol. 89, September 1980, pp. 101-107.

[92] Romslo, T. M. and S. F. Ravitz, "Arizona Manganese-Silver Ores", U. S. Bureau of Mines RI 4097 July 1947.

[93] Sano, K., "On the Reduction Equilibria of Cobaltous Chloride and Nickelous Chloride by Hydrogen", The Science Reports of the Tohoku University, Series I, Vol. XXXVII No. 1 (1953).

[94] Schafer, H. and K. Etzel, "Uber die Wanderung von Kupfer (I) Oxyd und von Kupfer in Temperaturgefalle, Z. Anorg. Allg. Chem. 291 (1957) 294.

[95] Scheiner, B. J., D. L. Pool and R. E. Lindstrom, "Recovery of Silver and Mercury from Mill Tailings by Electrooxidation", U. S. Bureau of Mines RI 7660.

[96] Scheiner, B. J., D. L. Pool, J. J. Sjoberg and R. E. Lindstrom, "Extraction of Silver from Refractory Ores", U.S. Bureau of Mines RI 7736, 1973.

[97] Stanczyk, M. H. and P. A. Bloom, "Copper Recovery from Segregation Flotation Concentrates by Ammoniacal-Ammonium Carbonate Leaching", U. S. Bureau of Mines RI 5826, 1961.

[98] Stetefeldt, C. A., "Russell's Improved Process for the Lixiviation of Silver Ores", Transactions AIME, Vol. 13, 1885.

[99] Sutulov, A., "Beneficiation of Chilean Oxidized Copper Ores by Segregation Process", World Mining, 15, No. 8, 1962, p. 28.

[100] Sutulov, A., "Proceso de Segregation en Beneficio de Cobrés Chilenos", Universidad de Concepción (1962) 66 pp.

[101] Sutulov, A., "Evaluating Copper Segregation Results by Screen Sizing Analysis", AIME Trans., 226, March, 1963, pp. 17-20.

[102] Trans. AIME, Vol. 17, p. 767, 1889 - The Goodale Process.

[103] Watson, B. N., "Large Low Grade Silver Deposits in N. America", World Mining, March 1977.

[104] Watson, B. N., "Bulk Tonnage, Low-Grade Silver Deposits - Update 1980", Precious Metals Symposium, Northern Nevada Section, SME/AIME with the Nevada Bureau of Mines and Geology and the Nevada Division of Mineral Resources, Sparks, Nevada, Nov. 17-19, 1980.

[105] Varley, T., E. P. Barett, C. C. Stevenson and R. H. Bradford, "The Chloride Volatilization Process of Ore Treatment", Bull. 211, U.S. Bureau of Mines (1923).

[106] Wyman, W. F. and S. F. Ravits, "Sulfur Dioxide Leaching Tests on Various Western Manganese Ores", U.S. Bureau of Mines RI 4077, 1947.

PATENTS, SEGREGATION PROCESS

[1] U.S. Patent 1,234,426, July 24, 1917, Vermaes Process.

[2] U.S. Patent 1,232,216, Aug. 3, 1917, Caron Process.

[3] Brit. Pat. No. 250,991, J. C. Moulden and B. Taplin, Oct. 29, 1924. Oxidized copper ore heated to a temperature below the melting point of copper, 500 to 700°C, with a reducing agent in the presence of a halogen or compound such as NaCl or $CaCl_2$ to effect reduction of copper oxide to metallic copper. Copper may be leached out after cooling or recovered by gravity concentration.

[4] Brit. Pat. No. 255,961, J. C. Moulden and B. Taplin, May 2, 1925. Heat treatment as in (1). Copper separated from treatment by froth flotation after grinding and screening, if required.

[5] Brit. Pat. No. 264,584, E. Edser and B. Taplin, Oct. 21, 1925. Oxidized ores containing lead, silver, antimony or bismuth can be similarly treated. If $CaCO_3$ is present in ore, pyrites and chlorides, such as those of Mn, Fe or Cu, may be added.

[6] Brit. Pat. No. 300,701, T. J. Taplin and B. Taplin, July 19, 1927. The procedure described in (1) is modified. Ore is partly reduced by reducing agent with halogen present and then treated to an oxidizing step for a short time, after which the reduction-halogenation is resumed. If native copper or copper sulfide is present in ore, the treatment may commence with an oxidizing step.

[7] Brit. Pat. No. 318,314, T. J. Taplin and B. Taplin, June 5, 1928. Gases evolved during segregation treatment can be recovered by scrubbing with water and returned to process. Gases can be scrubbed and treated with $CaCO_3$ or other material to form halides of Ca, Cu, Mg, Mn, Fe, etc.

[8] U.S. Pat. 1,679,337, J. C. Moulden and B. Taplin, July 31, 1928. Similar to (1). Temperature for segregation lower than melting point of copper, but sufficiently high to cause reaction (suitably 500° to 700°C). Mix ore with 2 - 2.5% of reducing agent and up to 0.5% NaCl or other halide like $CaCl_2$, $MgCl_2$, CaF_2, etc. Ore can be heated first in oxidizing atmosphere to oxidize any sulfides present. Recommends that soluble alkaline salts be removed prior to treatment or neutralized with H_2SO_4 or HCl. Chlorine or HCl gas can be used after ore-coal mixture is at operating temperature.

[9] Brit. Pat. No. 321,213, H. Lavers and B. Taplin, August 2, 1928. Adaptation of (2) in which a mixed sulfide-oxide ore is treated to segregation and then both metallic copper and residual sulfide froth floated together. Screening of coarse carbon with copper is also advocated.

[10] Brit. Pat. No. 348,024, T. J. Taplin, January 1, 1930. Ore subjected to sulfating roast or other material containing $CuSO_4$ can be treated by segregation process.

[11] U.S. Pat. No. 1,865,153, T. J. Taplin, June 28, 1932. Similar to (8). Found that segregated copper from copper sulfate is sulfided to a considerable extent, especially on its surface. This effect is believed to be due to the sulfur-bearing gases evolved. Mixed oxide-sulfide ores can be treated by first applying a sulfating roast.

[12] Brit. Pat. 387,713, T. J. Taplin, February 13, 1933. Ore is passed downward through a multiple hearth heating furnace wherein it is roasted and preheated to >650°C. The discharge falls into a heat insulated pit to which is fed continuously a mixture of NaCl and coal or coke, whereby Cu, Ag, Pb, Bi, and Sb reduced by segregation to charge.

[13] Brit. Pat. 389,865, F. B. Jones, March 27, 1933. Crushed ore mixed with carbonaceous material and NaCl is heated to 450° - 800°C in rotating tubular furnace from which air is excluded. Product is crushed and subjected to froth flotation to recover gold and copper.

[14] U.S. Pat. No. 3,148,974, C. Rampacek, September 15, 1964. A two-stage process is described consisting of an indirect-fired stainless steel preheat kiln and a direct-fired reactor to which the hot ore, reducing agent, and salt are added. The direct-firing of the reactor controls the temperature of the reactor furnace while simultaneously controlling the furnace atmosphere so that it is neither too highly oxidizing nor reducing. If copper sulfide or other sulfide minerals are present, these are oxidized in the preheat kiln. The air-natural gas ratio used is from about 1.2 to 1 times the ratio required to give complete combustion of the gas.

[15] U.S. Pat. No. 3,834,89, Eisel, J.A., and H. J. Heinen, Recovery of Gold - A variation of the segregation process using gaseous chlorine and a volatile compound of iron, aluminum or gallium, together with activated carbon. Recovery is by flotation.

RELATED PATENTS

[16] L. J. Bechaud and H. Hartjens. Segregation in the form of a sulphide (lead-copper) using pyrrhotite as reagent. U.S. Patent Application. Serial No. 781.442. December 19, 1958.

[17] H. Mino. Use of ammonium chloride. Jap. Patent Showa 39.30021, December 1964.

Chapter 4

SCALEUP EXPERIENCE IN GOLD–SILVER HEAP LEACHING

Daniel W. Kappes
Miller–Kappes Company
Reno, Nevada 89507

ABSTRACT

Testing ores for heap leach operations traditionally follows a progression of bottle roll tests, bucket and column leach tests, and very expensive large field leach tests. Enough data is now available to streamline the procedures. Data from five different operations shows that recoveries from large field leach tests can be predicted from the bucket tests. Initial recovery times in large field leaches are longer than in bucket tests – a typical response is eight days versus twenty days to the same recovery percentage; but the ultimate percent recoveries (after 60 days) are normally very close. Methods of mining and stacking can radically affect recoveries from clay–rich ores. Leaching of uncrushed ores provides special testing problems, since accurate samples smaller than several tons cannot be taken.

A precious metals heap leach is a fairly straightforward mechanism. Ore is stacked on an impervious pad and is sprinkled with a pH-controlled cyanide solution. The solution dissolves gold (and/or silver) from the rock and carries it down and out of the heap. Gold is removed from the solution and the solution is recycled.

The prediction of field heap behavior from laboratory tests should be relatively straightforward, provided that the laboratory test procedures are faithful mini-heap leaches. Detailed comparisons of field heap behavior versus lab behavior now exist for several ores. They indicate that field heap performance differs significantly, but predictably, from the laboratory tests. Specifically, field heaps reach the same level of recovery as the lab tests, but they take from three to six times as long to get there.

The data suggests that all production heap leach systems should be designed with flexibility to extend leach times for as long as the ore requires. Fortunately, Aesop's lesson that slow is not necessarily bad, is one of the guiding principles of heap leaching. The capability for unlimited leach times is inherent in the heap leaching process, both technically and economically. Surprisingly, this capability has been "designed out" of several operations in which the ore is stacked, leached, and removed on regular intervals. The data presented here, and evidence in the literature from other producing heap leaches, such as Cortez and Round Mountain[1], suggest that eliminating this flexibility greatly increases the investment risk.

Example 1. The Extremely Long Delay: A Clayey Ore

Figure 1 shows recovery curves in lab and field tests for two ores. Both ores leach quickly in the lab; but in the field they are at the extremes of heap behavior - one leaches very fast, the other very slow.

The laboratory test curves in Figure 1 were developed in typical small laboratory bucket leach tests, containing 20 Kg of ore in a 34 cm deep and 28 cm diameter column. A drawing of the test layout and details of the test procedure have been published previously.[2]

Ore A is a soft, almost pure kaolinite clay containing fine gold. Ore "lumps" are stable and permeable, but ore "fines" can form clay layers that are almost impermeable. To minimize the fines problem, the ore for the field heap was carefully blasted and was not crushed before leaching. The field heap was intentionally built in a manner that would result in delayed, steady gold recovery by using a conveyor stacker to construct the heap as a series of 1 meter high cones. Within each cone, the ore fines created an almost stagnant "pocket" about 0.6 meters in diameter. Around each cone the "throwing" action of the conveyor stacker created a layer of coarse rock which allowd free-flow of solutions downward through the heap.

As the curves for ore A in Figure 1 show, 60 percent recovery was achieved in lab tests in ten days, whereas the same recovery took 55 days in the field. It is not too surprising that the field and lab recovery curves for this ore are different since the tests differed in two significant ways:

1. There is no way to take an accurate sample of run-of-mine sized rock for small lab tests (these tests were run on ore crushed to 50 mm).
2. The throwing and segregating actions of the conveyor could not be duplicated in the lab.

The first factor appears to have been not too significant for this ore. Large production heaps have since been run on crushed ore with recovery curves that duplicate that from the uncrushed field test heap. The controlling factor appears to be heap permeability, not rock size.

The second factor is part of a problem common to all laboratory test columns, which is, that there is no way to duplicate the "structure" that is built into the field heap. Ore cannot be compacted the same way it is in the field, and ore segregation, which results from rocks rolling down the heap slopes, cannot be duplicated.

Example 2. The Short Delay

Ore B in Figure 1 is a mixture of permeable limestone and shale. Both the laboratory tests, and the field heap, were run on coarse-crushed ore.

The ore is hard enough so that ore fines do not clog the heap. As a result, it was possible to build the field heap as a vertical "stack" of 1.9 meter lifts, to a final height of 3.9 meters, using trucks which drove directly on top of each ore lift.

Laboratory tests were run in short bucket tests and in 2.5 meter high columns, and no differences in recovery times was noted. As in all other tests reported here, flowrates of leach solution in the lab and field tests were roughly identical.

Unlike ore A, the laboratory tests on ore B appeared to be faithful "mini-heaps", except for one aspect: the important, unduplicatable factor of heap structure. As Figure 1 shows, the laboratory tests achieved 85 percent recovery in eight days. The field heap reached this level of recovery in thirty days. Subsequent production heaps have all behaved identically to the first heap.

---

1. For a good general article and bibliography on heap leaching, see "Design Factors for Heap Leaching Operation" by George M. Potter, Mining Engineering, March 1981, pp 277 - 281.
2. "Leaching of Small Gold and Silver Deposits", by D.W. Kappes, in THE FUTURE OF SMALL-SCALE MINING, McGraw-Hill Publishing Co., 1981, pp 381 - 388. Edited by R.F. Meyer and J.F. Carmen. Copies available from the author.

Example 3. Ores That Require Crushing

The ore presented in Figure 2 is physically very different from either of the ores in Figure 1. This ore is a hard chert or jasperoid, in which very fine gold is apparently well disseminated. Recovery of the gold in laboratory tests is a straightforward function of the size to which the rock is crushed.

The field test curve shown in Figure 2 was generated from a 3.5 meter high field heap, with rock crushed to 8 mm. The ore in this heap was stacked using a conveyor stacker. There was no clay in the ore and no percolation problems could be detected in the field or in tall-column laboratory percolation tests.

As Figure 2 shows, the ultimate field recovery exactly matched the recoveries predicted from 13 cm high laboratort leach columns. However, as with ore B in Figure 1, there was a significant delay in field recoveries. Fifty percent recovery was achieved in eight days in the lab, versus twenty-seven days in the field.

The field delay appears to be a function of heap structure, not of heap height. This same ore was leached in a 6 meter high, 0.6 meter diameter laboratory column located at the USBM Reno Metallurgical Research Center. The recovery curves closely matched the small bucket tests rather than the field leaches.

Figure 3 shows a set of four gigantic columns, measuring 12.3 meters high, 1.25 meters diameter. The ore in these tests is "run-of-mine" size volcanic rock, with gold on fracture surfaces. The correlation for both ultimate recoveries and recovery times, between 13 cm high lab tests and these tall columns, has been very close.

Example 4. Flowrate and Rock Size Variations

It is easy, in a paper such as this, to present only the data which clearly makes the author's point. The ore in Figure 4 tends to muddle things a bit.

This ore is a soft porous limestone. It is physically similar to ore B in Figure 1, but it breaks differently. Simple bulldozer ripping results in "run-of-mine" ore which is 90 percent smaller thatn 1-inch. Because of its fine, earthy appearance, the heap was stacked (with ripped rock) using a conveyor stacker, in 1 meter high cones, to a final depth of 3.5 meters. Subsequent laboratory and field percolation tests showed that percolation rates were very satisfactory.

Three laboratory curves are shown on Figure 4. The effect of drastically lowering the flowrate had only a minor effect on recovery. Selected large rocks leached slowly and not very well, but the ore in the field heap was mostly small. The "prediction" from the lab tests would be for a 60 percent recovery after approximately 12 days of leaching. It took 50 days of leaching to achieve the same recovery from the field heap.

Example 5. Time Factors in Testing

Figure 5 makes a point somewhat removed from the general scheme of this paper. The figure shows laboratory test curves for three different ores. All three ores have approximately the same ultimate recoveries, near 60 percent. Had only short-term laboratory tests been run on these ores, however, the apparant picture would have been radically different. Recoveries after ten days ranged from 14 percent to 50 percent. A problem commonly faced by every testing laboratory is the "when can I have the answer?" syndrome, and a common response is to shorten the laboratory testing time.

Conclusion

In production heaps, allowing extra leaching time does more than insure against scaleup errors. At an ongoing project where the heaps are run countercurrently, the cost to leach "old" heaps is roughly $0.07 per ton per month. Significant additional cash flow can be generated from heaps producing only 0.5 percent additional recovery per month.

It appears that "time" must be added to a heap leach like "reagents" to a conventional mill; the approximate levels can be determined in the lab, but field operators must be given the flexibility to make final adjustments.

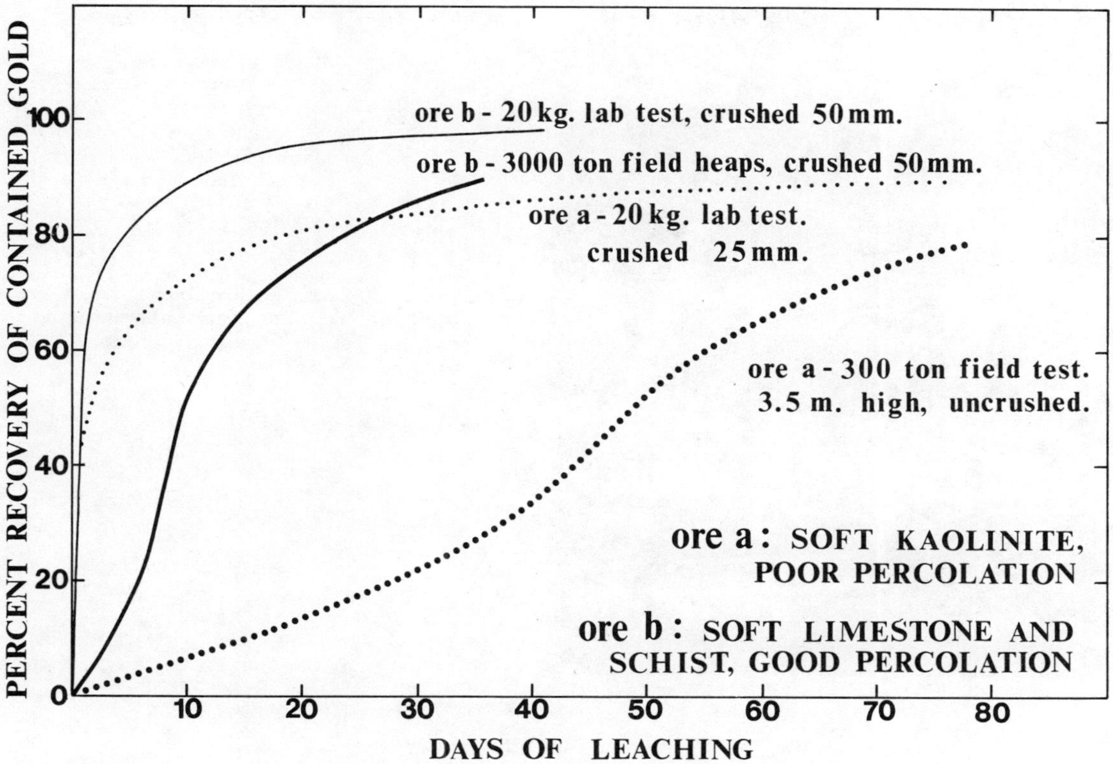

Figure 1: Comparison between lab and field leaches (two ores).

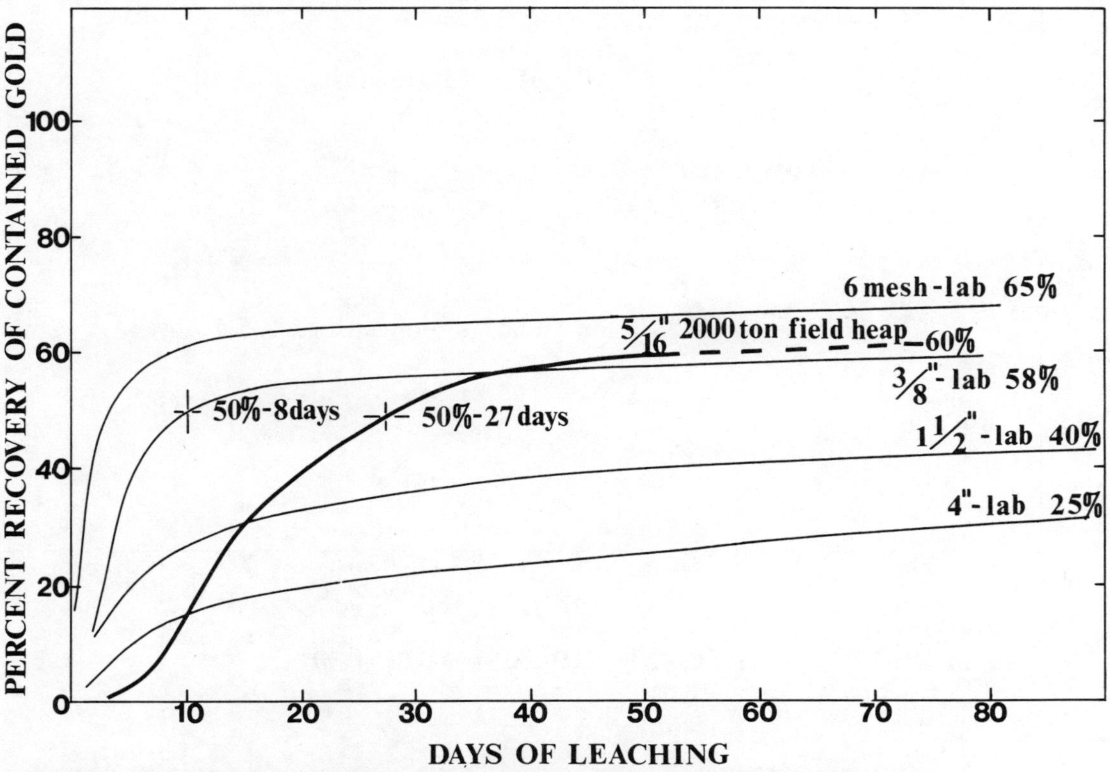

Figure 2: Comparison between lab and field leaches, crushed ore - no percolation problems.

FIGURE 3. Tall leach columns can provide data on scale-up behavior of chemically-reactive ores. These four leach columns, four stories high, hold 25 tons of ore each.

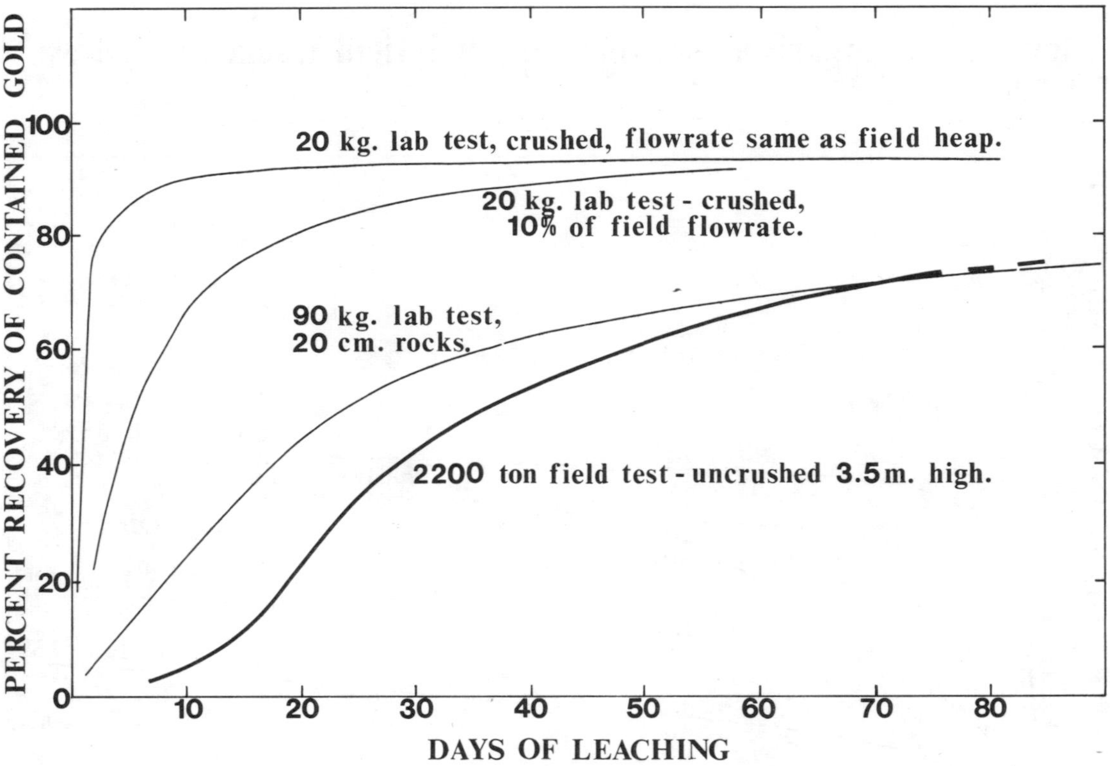

Figure 4: Effect of lab test variations.

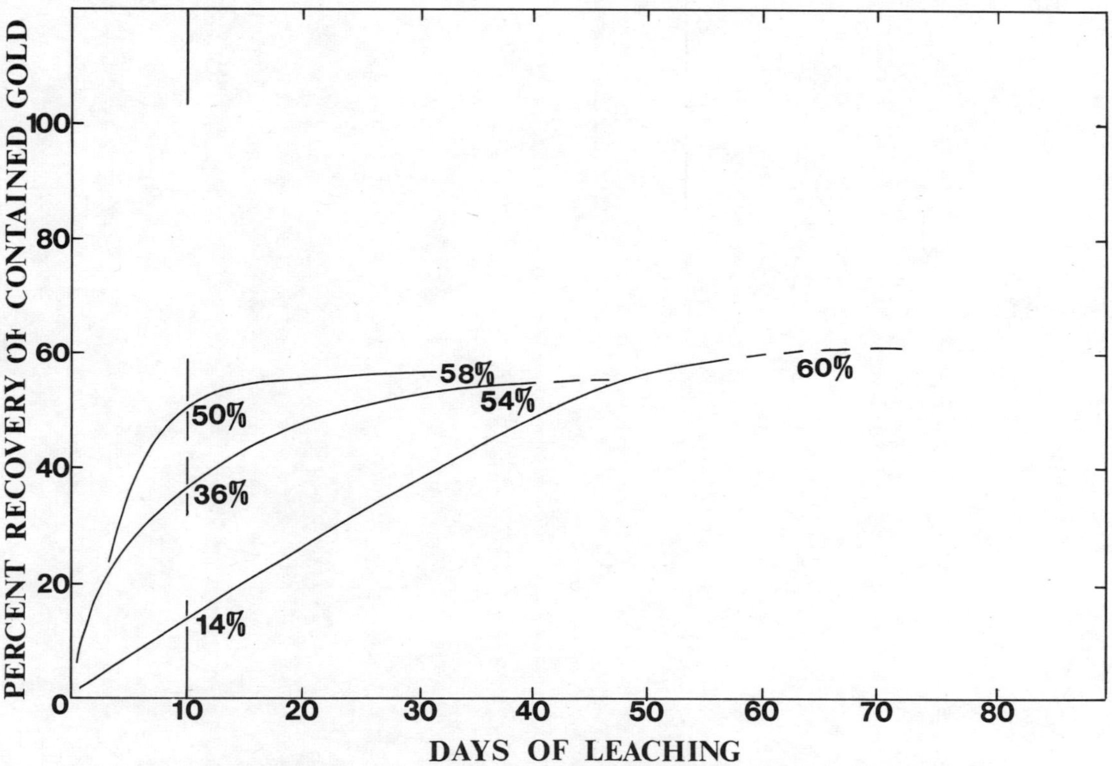

Figure 5: Leach time comparison laboratory small bucket tests, three ores, same conditions.

Chapter 5

HEAP LEACHING GOLD–SILVER ORES WITH POOR PERCOLATION CHARACTERISTICS

G. E. McClelland
S. D. Hill
Bureau of Mines
U. S. Department of the Interior
Reno, Nevada 89520

ABSTRACT

To maximize productivity and improve minerals processing technology, the U. S. Department of the Interior, Bureau of Mines, is investigating a particle agglomeration technique as a means of increasing the percolation rate of leaching solutions through heaps of clayey or finely crushed, low-grade, gold-silver ores. Bench-scale and pilot-scale experiments conducted on different ores indicated that the percolation rate of cyanide leaching solution was enhanced markedly by mixing the ore with a binder, such as lime or portland cement, moistening the mixture, then mechanically agglomerating and aging the feed prior to heap building and leaching. In addition, the rate of silver and gold extraction markedly increased without sacrificing total recovery of values. The use of concentrated cyanide solution instead of water during the agglomeration procedure substantially decreased the leaching time required to obtain maximum recovery. Results from bench-scale and pilot-scale experiments using the particle agglomeration technique are discussed.

INTRODUCTION

Exploration during the past few years has identified numerous low-grade, gold-silver deposits throughout the Western United States. Recent increases in gold and silver prices have generated a great deal of interest in processing these low-grade ores by heap leaching. Heap leaching with cyanide has potential application to many of these low-grade materials; however, if heap leaching is to be successful, the material must maintain good permeability after being stacked into heaps, so that the cyanide leaching solution percolates evenly throughout the heaps.

In an earlier publication,[1] the Bureau of Mines described the advantages of particle agglomeration as a pretreatment for certain materials which are difficult to treat by standard heap leaching. Excessive amounts of clay or fines generated during crushing prevented uniform percolation flow of the leaching solution. Laboratory tests showed that the percolation flow rates of cyanide leaching solutions were enhanced markedly by mixing the ore with controlled amounts of lime and water, agglomerating, and then aging the mixture prior to the heap leaching step.

The objective of the present investigation was to investigate portland cement (type II) as a binder in the particle agglomeration pretreatment and to transfer the technology to industry by cooperative pilot-scale studies in actual heap leaching operations.

LABORATORY EQUIPMENT AND MATERIAL

Investigations to determine the effectiveness of portland cement as a binder for the agglomeration of fine particles were carried out in column leaching experiments on 50-lb charges of material. A schematic of the laboratory apparatus is shown in Figure 1. The plexiglass columns were 5 ft high and had an inside diameter of 5.5 in. A removable porous support, 4 in. thick, was prepared from washed sand and gravel, which was enclosed in a 35-mesh nylon screen, and was placed at the bottom of the leaching column to prevent the ore from plugging the solution outlet. A 50-lb charge of material was placed on top of the porous support and gave a bed height of about 4 ft. A three-way plastic discharge valve was used to measure the flow rate and take solution samples. The pregnant cyanide solutions were pumped through three 1.5-in.-diameter columns in series, each of which contained 30 grams of 6x16 mesh coconut shell activated carbon which adsorbed the gold and silver. The barren solution was recycled to the top of the column of ore.

The clayey gold ore described in RI 8388 was also used in this laboratory investigation. A 39-in. rotating disk pelletizer was used to agglomerate the feed for the columns. Binder, dry ore, and a controlled amount of water or strong cyanide solution were mixed and agglomerated. The agglomerated material was aged prior to leaching. After curing, the feed was placed in the column and leaching started by pumping solution onto the top of the column.

INVESTIGATION OF PROCESS VARIABLES

Both lime and portland cement can be employed as binders for agglomerating fine and clayey particles. Portland cement additions of 5 to 10 lb per ton of clayey ore produced exceptionally stable agglomerates with high porosity. Binding properties of the portland cement were superior to those exhibited by lime. The agglomerates produced using cement as a binder endured percolation leaching using spray or flooding methods to simulate heap and vat leaching, respectively. No fines migration or solution channeling were observed. Therefore, research investigations concentrated on portland cement as the binder.

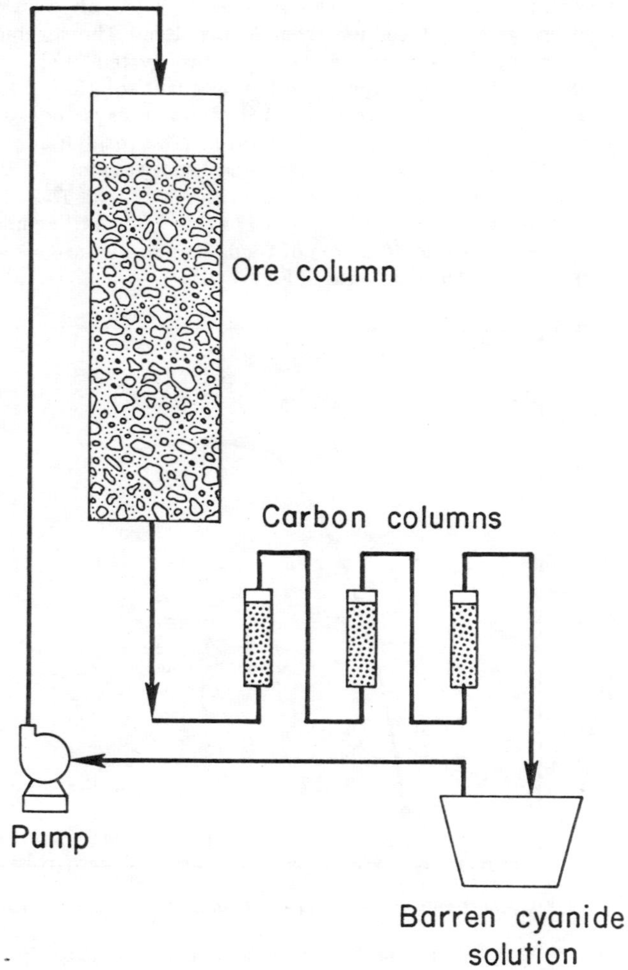

Figure 1. Flow Diagram for Simulating Heap Leaching Experiments

Three principal process variables were investigated and were 1) the amount of binder mixed with the dry feed, 2) the curing time required for the hydration of the calcium silicate bridges, and 3) the amount of moisture used to wet the dry mixture.

Effect of the Amount of Portland Cement Added on Percolation Flow Rate

Baseline Data. Initial column heap leaching tests were conducted on the clayey gold ore to obtain baseline data for comparison with data obtained from agglomerated feeds. Different amounts of cement were added to 50-lb charges of clayey gold ore crushed to a nominal 3/8-in. feed size, but the material was not agglomerated. The cement provided protective alkalinity and showed the effect of portland cement addition without agglomeration. The amount of cement was varied from 0 to 20 lb/ton of feed. After mixing the cement with the dry feed, the mixture was placed in the leaching column, and the downward percolation leach was started using 12 liters of solution containing 2 lb of sodium cyanide per ton. The leaching solution was recirculated through the system until no additional gold was detected in the pregnant solution. Flow rate measurements were made daily for five days and were averaged to determine the flow rate. The results showed that the percolation flow rate was independent of the amount of portland cement added to the dry feed. The flow rate remained constant at 0.12 gal/hr/ft^2 at cement additions from 0 to 20 lb/ton of feed. The results are shown graphically as baseline data in Figure 2.

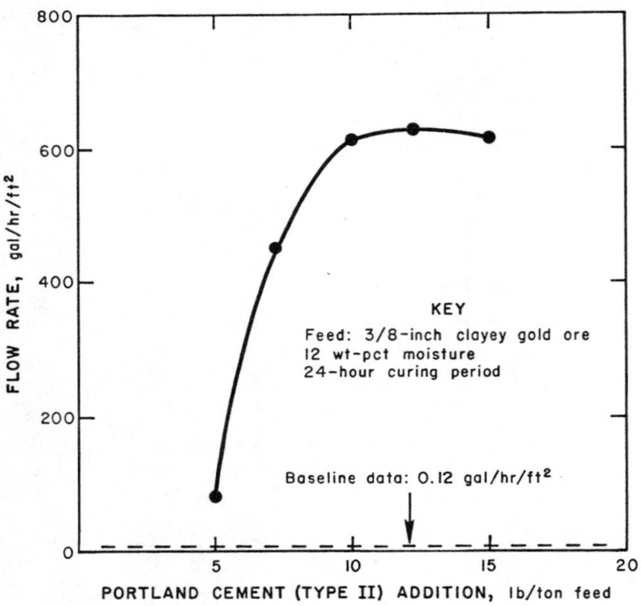

Figure 2. Effect of Binder Addition on Solution Flow Rate through Column

Effect of Portland Cement Addition on Agglomeration

Fifty-pound charges of the clayey gold ore crushed to a nominal 3/8-in. size were mixed with from 0 to 15 lb of cement per ton of feed. Water was added to the cement-ore mixture until the moisture content of the mixture on the 39-in. rotating disk pelletizer was 12 wt.%. The wetted mixture was tumbled on the pelletizer until the fine and/or clay material was agglomerated. Adequate mixing of the cement and water into the ore was important for successful agglomeration of the feed. During the agglomeration step, the fine particles adhered to the surface of the larger particles; thus, avoiding particle segregation. The coarse particles were coated by fine particles. Coating of the coarse particles did not affect precious metal recovery. The fines coating the larger particles were porous enough so that cyanide solution could penetrate and dissolve the micron size gold particles associated with the coarser material.

The agglomerated feed was placed in the leaching column and cured for 24 hours at ambient temperature. The column was capped to minimize drying of the pellets. An increase in agglomerate strength was observed if the cement-ore mixture remained moist during curing. If the pellets dried during curing, the hydrolysis reaction stopped and partial breakdown of the agglomerates occurred upon wetting.

After curing, downward percolation leaching of the agglomerated material was initiated as described above. During cyanidation, no additional base was required to maintain the leach solution pH at 11. Figure 2 shows that increasing the amount of cement up to 10 lb/ton of feed markedly improved percolation rates through the column of ore. Figure 2 also indicates that high flow rates are achieved by agglomerating the clayey gold ore with 7.5 to 10 lb of portland cement per ton of ore. This amount of cement will supply the protective alkalinity required during leaching under normal heap conditions. Maximum flow rate measurements were made after leaching was completed. The material in the column was flooded with leaching solution, and the rate at which the solution drained from the column was measured.

The high percolation flow rates obtained under flooded conditions would be impractical for solution pumping requirementes in an actual heap leaching operation, but the data demonstrate that very stable, porous agglomerates are produced and do not break down under exaggerated leaching conditions. The data also indicate that vat leaching techniques could be employed for leaching when portland cement is used as a binder.

Effect of Curing Time on Percolation Flow Rate

A series of column leaching experiments was conducted to determine the effect of curing time on the percolation flow rates. Fifty-pound charges of clayey gold ore were mixed with 10 lb of cement per ton of ore, wetted with 12 wt.% of water, and agglomerated on the disk pelletizer. The agglomerated charges were cured for 0, 2, 4, 8, 16, 24, and 36 hours at ambient temperature and in capped leach columns. Flow rates are presented in Figure 3.

The data show that the duration of the curing process is an important operating parameter. Percolation rates increased markedly up to eight hours of curing, but no further improvement occurred after eight hours. Curing the agglomerated feed for eight hours resulted in a flow rate of 600 gal/hr/ft^2 compared to the baseline flow rate of 0.1 gal/hr/ft^2, or a 6,000-fold increase in percolation. When lime was used as a binder, the best curing time was 24 hours.

Effect of Water Addition on Percolation Flow Rate

These experiments were performed on 50-lb charges mixed with 10 lb of cement per ton of feed. A 24-hr curing

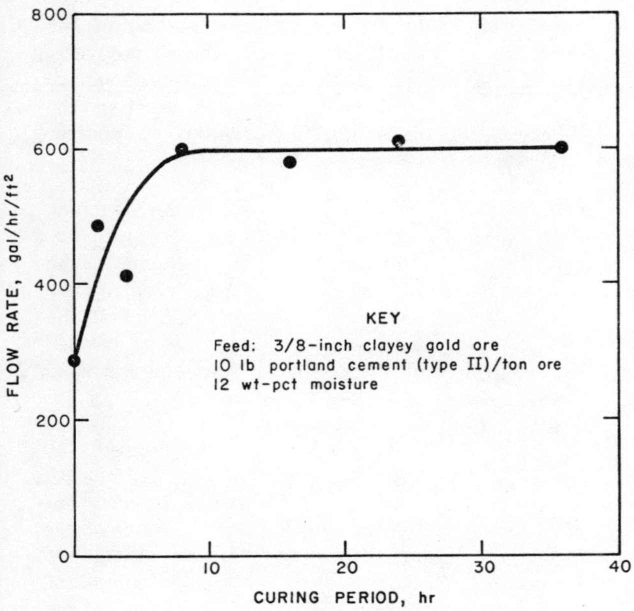

Figure 3. Effect of Curing Period on Solution Flow Rate through Column

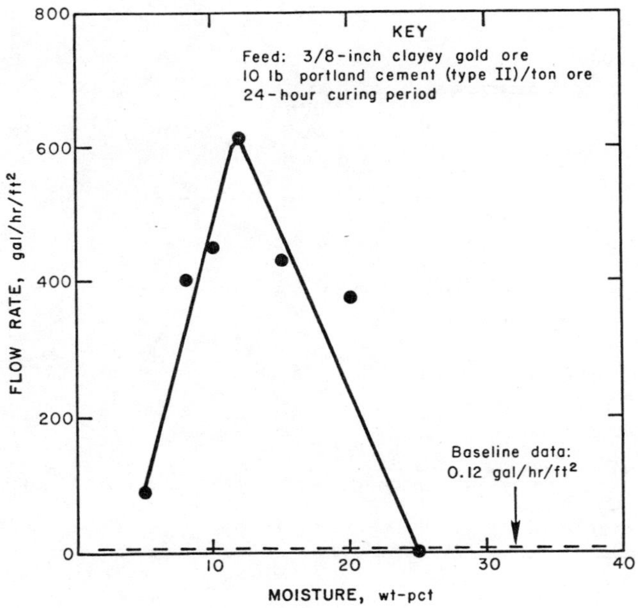

Figure 4. Effect of Moisture Content on Solution Flow Rate through Column

period was used as an experimental convenience even though only eight hours were needed to achieve full curing. The results are shown in Figure 4. Solution flow rate through the column of agglomerated ore increased considerably with increasing moisture content, attained a maximum of about 600 gal/hr/ft^2 at 12 wt.% moisture, and then rapidly decreased to less than baseline at 25 wt.% moisture. These data indicate that the permeability of the agglomerated feed is greatly dependent upon the amount of water employed in the agglomeration step. If too much water is added, the feed-cement mixture becomes a mass of mud, and does not produce stable, porous agglomerates.

Summary

The best combination of cement, water, and curing time determined experimentally for the clayey gold ore is summarized in Table 1. Agglomeration pretreatment

TABLE 1
SUMMARY OF EXPERIMENTAL RESULTS

Best pretreatment conditions:	
Portland cement, lb/ton ore	7.5
Moisture, wt.%	8-20
Curing time, hour	8
Percolation rates:	
Baseline, no pretreatment, gal/hr/ft^2	0.1*
With optimum pretreatment, gal/hr/ft^2	610.0*
Gold extraction, %	70

* Percolation rate measurements were taken under flooded column conditions.

increased the percolation rate from 0.1 to 600 gal/hr/ft^2 of cross sectional area. Gold extraction was determined during the simulated heap leaching sequence. The 70% extraction obtained was almost identical to that obtained by agitation bottle leaching tests of 3/8-in. feed material.

Effect of Wetting with Cyanide Solution on Leaching Period

The effect of wetting the cement-ore mixture with concentrated cyanide solution instead of water during the agglomeration step was studied. By adding the required sodium cyanide to the feed during agglomeration, the leaching period required for maximum precious metal extraction might be decreased because the cyanide would be reacting with the gold during the curing period. Subsequent percolation leaching with water would recover the gold values.

A series of three column leaching experiments using 50-lb charges of the minus 3/8-in. clayey gold ore was conducted. Baseline-unagglomerated leaching, particle agglomeration with cement and water, and particle agglomeration with cement and cyanide solution were evaluated. The baseline test was conducted by placing the dry ore into the column and leaching with 2 lb of NaCN per ton of solution or 1 lb of NaCN per ton of ore. Sodium hydroxide was used for protective alkalinity. The cement-water agglomeration experiment was performed by mixing 10 lb of portland cement per ton of dry ore and wetting with water to bring the final moisture content of the feed to 12 wt.%, agglomerating the mixture on a rotating disk pelletizer, and aging the agglomerated material for 24 hours prior to leaching with 2 lb of NaCN per ton of solution. The cement-cyanide agglomerates were prepared in the same manner as described above except that the feed was wetted with a cyanide solution containing 8.6 lb of NaCN per ton of solution, and leaching was started with water. Results from these experiments are shown in Table 2.

TABLE 2
COLUMN LEACHING DATA

Pretreatment	Calculated head, oz/ton Au	Ag	Cement, lb/ton ore	Cyanide, lb/ton solution	Moisture, wt.%	Percolation Rate, gal/hr/ft^2	Leaching Period, days	Cyanide Consumption, lb/ton ore	Recovery, % Au	Ag
None	0.09	0.47	—	2	—	0.1	26	1.0	68.1	15.6
Particle agglomeration with cement and water	0.10	0.30	10	2	12.0	610.0	9	0.8	70.3	26.7
Particle agglomeration with cement and cyanide solution	0.10	0.40	10	8.6*	12.0	600.0	5	0.6	73.6	31.0

* Based on cyanide concentration of the pretreatment solution, the total amount of cyanide added per ton of ore was the same for all tests.

The data from these experiments show that agglomerating the ore with cement and strong cyanide solution decreased the leaching time for maximum gold extraction from 26 days for conventional heap leaching to five days for agglomeration heap leaching. Agglomerating the ore with cyanide solution decreased cyanide consumption.

PILOT SCALE TESTING OF AGGLOMERATION TECHNOLOGY

Several pilot scale tests employing particle agglomeration technology were conducted by private operators interested in applying the technique to their precious metal bearing material. Bureau of Mines personnel acted as advisors. Three pilot scale experiments will be described in this paper and were selected because of the diversity of the materials and methods of mechanical agglomeration employed. Feed materials used in the pilot scale agglomeration heap leaching tests included: 1) a gold ore containing few fines and no clays, 2) a clayey silver mine waste material, and 3) a silver-bearing mill tailings.

Gold Ore

An oxidized disseminated gold ore from an open pit mine located near Ely, Nevada, was employed in the first pilot scale experiment. The ore contained no clays and only 15 wt.% minus 20 mesh material when crushed to 80% minus 3/8 in. Although the ore did not require agglomeration pretreatment to ensure adequate percolation rates, the operators were interested in the effectiveness of particle agglomeration for controlling particle segregation during stacking and leaching of the heap. They were also interested in decreasing channeling of solution during heap leaching.

Two pilot scale tests were performed. The first test was made on 5,000 tons of feed material crushed to 80% minus 3/8 in. and agglomerated with water to control the fines while the heap was being stacked.

The feed was prepared by wetting the dry ore with water at the discharge end of the crusher. Partial agglomeration of fines occurred by mixing with the coarser ore as it rolled down the sides of the stockpile. Additional agglomeration of fines occurred as the wetted feed was pushed over the edge of the heap and rolled down the heap's sides. The feed material was carefully stacked on the heap in a manner described by Chamberlin.[2]

Leaching was accomplished by sprinkling alkaline cyanide solution containing 0.5 lb of NaCN per ton over the heap surface. The application rate of cyanide solution was 0.18 gal/hr/ft^2 of heap area. The leaching solution percolated through the heap and was collected on a water-impervious pad which drained into a plastic-lined pregnant solution pond. The pregnant solution from the pond was pumped through a series of columns containing coconut shell activated carbon which adsorbed the gold values. The barren solution was pumped to a plastic-lined pond for recycle to the heap. Makeup reagents were added in the barren solution pond. Pertinent leaching data obtained are shown in Table 3.

The second test was conducted on 2,500 tons of ore crushed to 80% minus 3/8 in. The ore was mixed with 10 lb of portland cement per ton of dry ore, and wetted with a solution containing all the NaCN required for the leach. The agglomerated material was aged on the heap during heap construction.

The feed material was transported with a front-end loader from the crushed ore stockpile to a 9-ton weighing hopper. Cement was added to the ore as it was conveyed from the hopper to a conventional 14-yard-capacity concrete mixer. Cyanide solution was introduced into the mixer in three increments. The final moisture content of the 9-ton batches of agglomerated ore was between 7 and 10 wt.%. Agglomeration of the ore occurred during the transport of the treated ore to a stockpile area. The agglomerated feed was transferred from the stockpile by a front-end loader and dumped over the edge of the heap. The same leaching/carbon adsorption procedure employed in the first test heap was used except that no makeup reagents were required. Pertinent leaching data comparing both test heaps are shown in Table 3.

TABLE 3
LEACHING DATA FROM TWO PILOT-SCALE LEACHING EXPERIMENTES

Pretreatment	Calculated head oz/ton Au	Cement, lb/ton ore	Cyanide, lb/ton solution	Moisture, wt.%	Solution Application Rate, $gal/hr/ft^2$	Leaching Period, days	Cyanide Consumption, lb/ton ore	Au Recovery, %
Particle agglomeration with water alone	0.17	—	0.5	7-10	0.18	60	0.3	88
Particle agglomeration with cement and cyanide solution	0.17	10	8.0*	7-10	0.18	40	0.3	85

* Based on cyanide concentration of the pretreatment solution, the total amount of cyanide added per ton of ore was the same for all tests.

required to obtain the best gold recovery. Gold recoveries were the same as those obtained in small scale column percolation experiments. Particle agglomeration with water or with cement and cyanide solution was effective in controlling the migration of fines during stacking and leaching of the heap and in decreasing solution channeling during leaching.

Silver Mine Waste Material

The silver mine waste from the old Packard mine workings near Lovelock, Nevada, contained considerable fine material and required particle agglomeration in order to obtain an adequate flow rate through the stacked heap. Before applying particle agglomeration, a 30,00-ton heap of mine run waste was constructed on a water-impervious pad. When alkaline cyanide solution was applied to the heap, it ran off the sides of the heap rather than percolating through the bed of material. No silver was recovered from the material using conventional heap leaching. Several laboratory scale column leaching experiments were performed by the Bureau of Mines to evaluate the feasibility of applying particle agglomeration techniques to the material. Since results from the laboratory experiments were encouraging, the operators of the property conducted pilot scale particle agglomeration/heap leaching tests. Three 300-ton heaps were constructed and leached to determine the best feed size and leaching conditions. These tests showed that the feed had to be crushed to 80% minus 9/16 in. to liberate the silver values. Leaching with dilute cyanide solution for six days at a flow rate of 9 $gal/hr/ft^2$ of heap area achieved a 65% silver recovery.

Using the agglomeration and leaching conditions determined by the above tests, the operators constructed and leached a 1,000-ton heap. Dry material from the original 30,000-ton heap was transported to a hopper where it was conveyed to a two-stage roll crusher. The nominal 9/16-in. feed from the crusher was conveyed to the agglomerator.

Agglomeration was accomplished by wetting the crushed feed with a cement slurry. Ten pounds of cement per ton of feed were added, and the final moisture content of the agglomerated feed was maintained between 9 and 11 wt.%. The moistened feed tumbled down a vibrating inclined chute, which was 4 ft wide and 8 ft long and contained several "stair steps" to effect agglomeration.

The operators experienced difficulty in applying the cement in a slurry. Insufficient agitation in the cement-water mixing tank caused the cement to settle and the feed was wetted with only one quarter of the cement required for adequate agglomeration.

Another problem encountered in adding the cement as a slurry was the effective life of the wet cement. The cement hydrolyzes rapidly so that it is ineffective as a binder in less than five hours. However, even with the problems experienced by adding the cement as a slurry, the pilot-scale experiment was successful. The agglomerated feed was conveyed to a stockpile and later transferred with a front-end loader to a water-impervious leaching pad.

The 1,000-ton agglomerated heap was leached by sprinkling dilute cyanide solution over the heap at a rate of 9 $gal/hr/ft^2$ of heap area. The pregnant solution was collected on the water-impervious pad and drained into a plastic-lined pond. A Merrill-Crowe zinc precipitation method was used to recover the silver from the pregnant solution. The resultant barren solution was recycled to the heap. Sixty percent of the silver in the ore was recovered in six days of leaching.

Particle agglomeration with cement and water was a viable processing technique for this material, whereas conventional heap leaching was unsuccessful. The operators plan to incorporate this technology into their commercial operation. However, dry cement will be mixed into the feed, and the mixture moistened with a solution containing all the NaCN required to dissolve the contained silver.

Silver Mill Tailing

The third pilot scale test was performed on a silver mill tailing containing 2.5 oz of silver per ton. The tailings were from an old cyanide mill located near Virginia City, Nevada. The tailings which were 65% minus 200 mesh had been exposed to weathering for several decades.

Laboratory-scale column leaching experiments indicated that the material could not be leached using conventional heap leaching methods because the material was too fine. The tailings were too low grade to warrant construction of a conventional agitation cyanide plant. Several laboratory-scale agglomeration experiments were made and gave encouraging results. The operators decided to conduct a small pilot scale test to determine the effectiveness of agglomeration technology for recovering the contained silver values. Forty tons of silver-bearing tailings were mixed with 30 lb of portland cement per ton of feed, wetted with a strong cyanide solution so that the final moisture content of the feed was 22 wt.%, agglomerated in a rotating drum and then aged on the heap for several days prior to percolation leaching.

The feed was placed into a hopper and screw-fed to a rotating drum. The drum was 42 in. in diameter and 7 ft long. Dry cement was mixed with the tailings at the discharge end of the screw feeder. The mixture was moistened with cyanide solution applied through three spray nozzles mounted inside the balling drum. A scraper was used inside the drum to prevent material buildup on the walls of the drum. The agglomerates from the drum were conveyed to the heap.

Leaching was initiated by spraying water over the surface of the heap. The pregnant solution was collected on a water-impervious pad and drained into a pregnant solution reservoir. The pregnant solution was pumped upward through a series of five columns containing coconut shell activated carbon which adsorbed the silver values. The resultant barren solution was recycled to the heap. A silver recovery of 78% was obtained in five days of leaching at a percolation rate of 12 gal/hr/ft^2 of heap area. A silver recovery of 81% was obtained in the laboratory agglomeration experiments.

Particle agglomeration with cement and cyanide solution again demonstrated its effectiveness as a processing technique for materials that possess poor percolation characteristics. The quantity of cement used in the pilot scale test was more than required because laboratory results showed that 15 lb of cement per ton of feed was sufficient for adequate agglomeration. The operators of this property intend to use particle agglomeration/heap leaching to process the silver-bearing mill tailings.

POTENTIAL BENEFITS USING AGGLOMERATION TECHNIQUES

Several potential benefits can be attributed to particle agglomeration pretreatment of ores and tailings before heap leaching:

- Clayey and finely crushed ores can be treated by low-cost heap leaching procedures. This permits treatment of large low-grade ore bodies, many of which are presently considered submarginal for cyanide mill constsruction.
- Increased gold recovery from an ore can be obtained because additional gold may be liberated by finer crushing without encountering particle segregation during preparation of ore heaps. Particle segregation in unagglomerated materials can cause localized accumulation of fines that inhibit flow of leaching solutions.
- Percolation rates are increased, thus decreasing the required leaching cycle. This means that the mine capacity can be increased without increasing capital cost for pad preparation.
- The highly porous nature of the agglomerates permits the heaps to "breathe," thereby providing the oxygen necessary for the gold dissolution. The height of the heap can be increased so that the pad preparation cost per ton of ore processed is less, and land area is more effectively utilized.
- Favorable environmental conditions are realized in that the highly porous structure of the heap permits efficient washing of residual cyanide from processing heaps. The stable agglomerates minimize dusting problems when the heaps are abandoned.
- Precious metal concentration in solutions exiting from the heaps are greater. Efficiency of gold-silver recovery from these solutions by precipitation or adsorption on activated carbon is enhanced.

CONCLUSIONS

Solution flow rates through ore heaps are increased by agglomerating the fine particles in the ore with portland cement and water or cyanide solution, then curing. The agglomeration procedure eliminates the problem of particle segregation during the stacking of crushed ore and results in more rapid recovery of the precious metal values, smaller soluble gold and silver losses, and more thorough washing of the residual free cyanide in leached heaps. By improving percolation leaching, this technique has the potential of helping to solve environmental problems associated with tailings disposal and of making an important economic impact on gold-silver recovery operations.

REFERENCES

1. Heinen, H. J., McClelland, G. E., and Lindstrom, R. E., "Enhancing Percolation Rates in Heap Leaching of Gold-Silver Ores," BuMines RI 8388, 1979, 20 pp.

2. Chamberlin, P. D., "Heap Leaching and Pilot Testing of Gold and Silver Ores," Mining Cong. J., Vol. 67, No. 4, April, 1981, pp. 47-52.

Chapter 6

LEACHING PRACTICES AT SMOKY VALLEY MINE

C. Arthur Lefler
Smoky Valley Mining Division
Copper Range Co.
Round Mountain, Nevada 89045

ABSTRACT

The hydrometallurgical processes for recovery of gold and silver in use at Round Mountain, Nevada and the practical aspects of the operation are discussed. Included are heap leaching by cyanidation, upgrading of solutions by adsorption of values on activated carbon and desorption (stripping) for electrowinning and subsequent refining steps to produce dore' metal.

INTRODUCTION

The Smoky Valley Mine is jointly owned by Louisiana Land and Exploration Company, Felmont Oil Corporation and Case Pomeroy Company. Copper Range Company, a subsidiary of L.L. & E. is the operator of the mine.

The mine is located about 60 miles (100 km) north of Tonopah, Nevada on Highway 376 (formerly 8A) in the Big Smoky Valley on the western flank of the Toquima Mountain Range in the shadow of Mount Jefferson.

The mine is at an elevation of about 6100 ft. (1860m) and is situated on the southwestern flank of Round Mountain, the hill for which the town of Round Mountain is named. The ore is a rhyolitic ignimbrite with coarse gold occurring along fractures and with fine microscopic gold disseminated in the rock mass. Minor silver is associated with gold.

Annual mining is planned at 5.8 million tons (5.3 million tonnes) per year with 2.2 million tons (2 million tonnes) as ore (7,000 tons or 6,400 tonnes per day).

Material Handling

After drilling and blasting the ore is loaded on fifty (50) ton (45 tonnes) trucks with rubber tired loaders or an electric power shovel and transported to an open circuit crushing system where it is reduced in size to a normal - 1/2 inch (1.25cm) product. The crushed material is belt conveyed to a storage pile and is then withdrawn as required and conveyed to a truck loading bin. The crushed ore is transported to the leaching pad area by the 50 ton (45 tonnes) trucks and stacked to a height of 13 feet (4m) using Loaders.

The material after leaching, washing and draining, is excavated from the leach pads with a loader and transported by trucks to an adjacent dump area.

Leaching Pad Construction

The leaching pad is constructed of Hydraulic asphalt concrete over a carefully compacted base. The build-up of the pad includes a two inch (5cm) layer of asphalt which is covered with an asphalt rubber membrane and two additional 2½ inch (6.4 cm) layers of asphalt concrete. This provides an impermeable barrier to the downward percolating solutions.

The leach pad is 2525 feet (770m) long and 282 feet (86m) wide. The width includes a common drainage ditch which collects all effluent solutions and conducts them to a pregnant solution pump sump.

Application of Leaching Solution to Heaps

Current practice for application of leach solution includes use of polyethylene piping for conveying barren solution from the plant effluent sump to the leach pads and the use of removable PVC piping for the spray distribution systems on the heap surface.

The barren solution is applied to the heaps by spraying - using "Bagdad Wigglers" 12" (30.5cm) lengths of 1/4" (0.64cm) surgical rubber tubing. The "Wigglers" are arranged in a 35 ft. x 35 ft. (10.7 x 10.7m) square pattern on the surface of the heaps with output of each spray at 5 gpm. (.3 L/s). The current application rate is .004 gpm/ft^2. (.0027 L/s/m^2). We are planning to increase the application rate to .005 gpm/ft^2 (.0034 L/sec/m^2) to improve coverage. We are now distributing 2200 gpm (.139m^3/s) of solution on the heaps but will increase this shortly to 3000 gpm. (.189m^{3}s).

Heap Leaching Metallurgy

Metallurgical treatment consists of application of an aqueous solution of sodium cyanide and sodium hydroxide to the ore heaps to dissolve the gold and silver values.

Leaching solution concentrations to the heap are maintained nominally at 0.05% cyanide and 0.03% NaOH and solution PH is maintained at 11.0.

The leaching cycle is nominally 48 days distributed as follows:

Leaching Time	40 days
Washing Time	2 days
Draining Time	2 days
Loading, Unloading and Open	4 days

The leaching solutions are in closed circuit and the only make-up water is for evaporation and moisture in leached tails. The usual source of make-up in the system is the wash water which essentially balances with the losses for evaporation and tailings moisture. All solution from the leach pads is collected in a common ditch and by gravity to a pregnant solution sump.

The composition of pregnant solution is commonly from .020 to .030 troy oz Au/ton (.7 to 1.0g Au/tonne) of solution and from .4 to .6 lbs (181 to 272g) cyanide/ton of solution at a PH of 9.5 to 10.0.

Process Plant

Recovery of gold and silver from solution is by activated carbon adsorption, desorption and then electrowinning and fire refining. The final product is dore' metal typically containing 65% gold 34% silver, and minor amounts of copper, nickel and iron.

Adsorption Section

Pregnant solution at the rate of about 1600 gpm. (.101m^3/s) is pumped from the pregnant solution sump to an elevated plant inlet screening station. The solution then flows by gravity through a series of five 12 ft. (3.65m) diameter x 8 ft. (2.44m) deep open top adsorption reactor vessels which contain activated carbon. Flow through each tank is upward through the expanded carbon bed. Each reactor contains four tons (3.6 tonnes) of 12 x 30 mesh coconut carbon.

Carbon transfers are made counter to the solution flow in one ton (.9 tonnes) batches each day into the reactor system, between reactors and from the reactor system.

The carbon transfers are accomplished by eduction using barren solution as the transfer medium. Carbon is loaded during the five stages of adsorption to approximately 200 troy oz Au/ton (6900g Au/tonne) of carbon.

Desorption (Stripping) Section

One ton (.9 tonne) of carbon is withdrawn from the first reactor and passed daily to the desorption or stripping section which consists of a conditioning tank and three desorption tanks. The carbon is placed in the conditioning tank and raised to about 180°F (82°C) in the stripping solution and held for one day. It is then transferred to one of the three desorption vessels and leached for three days with sodium cyanide and sodium hydroxide solutions at a temperature of 180°F (82°C). The stripping solution at the rate of about 18 gpm. (1.14 L/sec) is passed through the three vessels in series fashion - into the oldest and out of the newest and the stripping effluent, containing about 2 oz Au/ton (68g Au/tonne) of solution, is fed continuously to the electrowinning section.

Electrowinning

The electrowinning section consists of three 2' x 3' x 4' (.6m x .9m x 1.2m) cells with parallel solution flows of 6 gpm (0.4 L/sec). Each cell has three 2' x 3' (.6m x .9m) stainless steel anodes and six steel wool cathodes. The cells operate at 3 volts and 140 amps or about 4 amperes per square foot (43 A/m^2). Usual pulling schedule for each cell is as follows: The two cathodes closest to input side are pulled each week, the remaining four are shifted toward the input side of the cell and the new steel wool cathodes are then put into the spaces left at the output side of the cell. The dore' precipitates with residual iron from the steel wool cathodes are removed and dried to prepare for fire refining.

Refining

The dried precipitate from the electrowinning cells is weighed and nitre, borax and silica reagents are added in the proper proportions for the refining furnace charge. The furnace is then charged and a temperature of 2100°F (1150°C) is obtained and held for 1.5 hrs. Each charge produces 200 to 400 oz (6200 to 12400g) dore' metal buttons. The buttons are then accumulated and melted in the casting furnace and dore' bars of 1000 to 1150 oz (31100 to 35800g) are cast.

Slag from button production is pulverized and then screened to remove the coarse trapped dore'. The screen undersize is tabled to remove fine trapped dore'. These two products are returned to subsequent melts.

Leach Solution Heating

Temperatures during the winter in the Big Smoky Valley commonly fall well below freezing causing shutdown of the leaching heaps for the most of December, January and February each year. In order to mitigate losses due to freeze ups, submerged combustion burners capable of providing 25 million BTU's per hour ($26.4x10^9$ joules/hr) to the leaching solution have been installed and are used as necessary to allow operation of the leaching system when temperatures are as low as 15°F (-10°C). Even in the worse cases where shutdown cannot be prevented, resumption of production is much more rapid.

Carbon Treatment

We have experienced a considerable build-up in the carbon of calcium carbonate. The effect of the build-up caused a decrease in adsorption efficiency resulting in high plant tails. We are now washing the carbon with nitric acid and are operating at former levels of recovery. Tails from the plant are at .002 troy oz per ton (.07g/tonne) of solution. The technique employed is to place one ton (.9 tonnes) of stripped carbon, after kiln reactivation, in a fibreglass vessel and acid solution (10% HNO_3) is circulated through the carbon until a constant PH of 6 is achieved. Carbon is then washed and made basic with sodium hydroxide and reinserted into the adsorption circuit. Since we reactivate one ton (.9 tonnes) of carbon per day, we wash and hold the carbon for the 24 hour period.

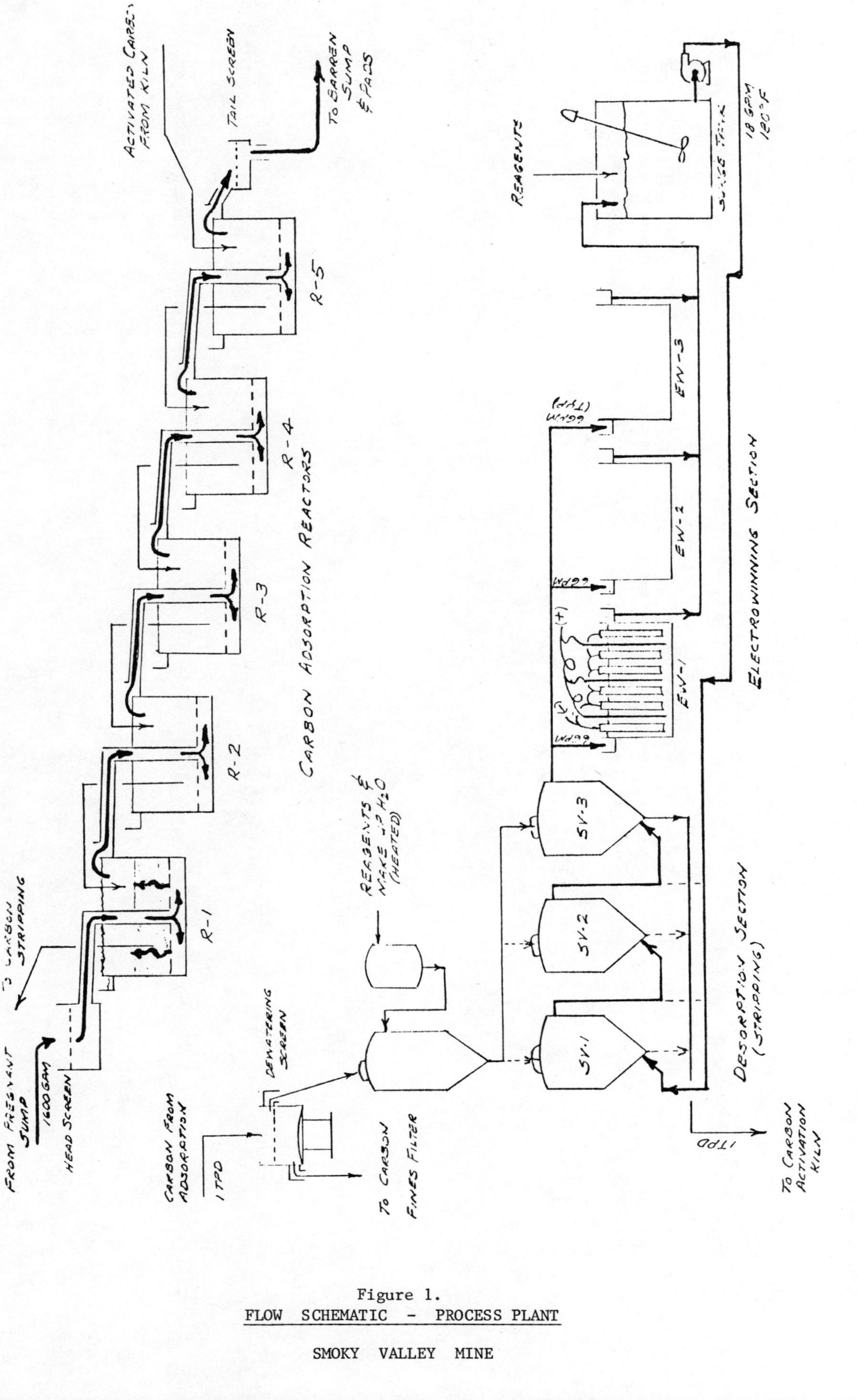

Figure 1.
FLOW SCHEMATIC - PROCESS PLANT

SMOKY VALLEY MINE

Chapter 7

UNITIZED, PORTABLE FLOTATION–LEACH PLANTS FOR TREATING PARTIALLY OXIDIZED GOLD AND SILVER ORES

Paul H. Johnson, Ph.D.
Resource Engineering & Development, Inc.
7117 South, 400 West, Unit 3
Midvale, Utah 84047

ABSTRACT

Resource Engineering and Development, Inc. is desigining and building a portable 150 to 225 tpd, flotation-cyanide leach plant with unitized, skid-mounted components. This plant design, with its portability, low capital investment, and high recovery of silver and gold from many partially oxidized ores, is well suited to treat many small, high-grade deposits, old waste dumps, and old mill tailing piles. A description is given of flotation and/or cyanide leach tests run on 18 partially oxidized, newly mined ores; waste dump materials; placer slimes; and old mill tailings.

INTRODUCTION

The large increase in precious metal prices over the past few years has generated considerable interest in the milling or remilling of many old, partially oxidized, mine dumps, tailing piles, stope fill, and ores mined from oxidized surface zones.

The present technology is often inadequate to treat these oxidized gold and silver ores. In several instances, heap or pad leaching plants have failed because of low metal recoveries and the necessity of winter shutdowns. Many of these ores respond poorly to treatment with either flotation or cyanide leaching alone.

The size and grade of many of these types of deposits will usually not warrant the expenditure of capital needed for a conventional, stationary, flotation or cyanide leach plant.

A simplified, unitized, portable, flotation-cyanide leach plant may be the solution to these problems. Considerable testing has indicated that a combination flotation-leach plant will successfully recover most of the silver and gold values from many of these oxidized ores.

Resource Engineering and Development, Inc. (R.E.D.) is currently designing and building a 150- to 225-ton-per-day (tonnage dependent on ore feed size) unitized, portable, flotation-leach plant for Silver Bullion Milling and Refining, Joint Venture. The first operating site will be near Hailey, Idaho, where the plant will be used to treat old flotation mill tailings, old minus ¼-in. jig mill tailings and several old mine dumps (95% minus 2-in.). Eventually, the plant may be used to treat other company-owned or custom ores in the Central Idaho area or it will be moved to another site(s).

UNITIZED, PORTABLE, FLOTATION-LEACH PLANT DESIGN

The general concept of the unitized, portable, flotation-leach (UPFL) plant is one in which both flotation and carbon-in-pulp cyanide leach circuits are incorporated into a greatly simplified mill design of commercial size, built in unitized, skid-mounted modules.

The flow sheet of a 150- to 225-ton-per-day plant of this type is as shown in Figure 1. In this plant design, prescreened, minus 2-in. ore will be periodically loaded into a 35-ton ore storage bin, from which it will be fed at a constant rate to a 7 ft x 48 in. Hardinge conical ball mill with an internal grate. A grizzly (6-in. opening) on the storage bin and a single, close-set jaw crusher might be used in the circuit if coarse, hard ores were to be fed to the plant. The ball mill discharge will be classified by a cyclone which produces an overflow product with a pulp density of about 40 to 45% solids and a product size of about 95% minus 65 mesh. Flotation reagents and recirculated solutions from the tailing pond will be added to the grinding circuit. The cyclone overflow will go to a DSM stationary screen with a 28 mesh deck to remove oversized ore particles and small wood pieces from the plant feed slurry, and then to a bank of rougher flotation cells with a 10-min. retention time. The flotation concentrates will be cleaned once or twice, as required, and then, since they are present as a small tonnage per day, sent to a decant pond and a concrete pad for air drying prior to shipment to a custom smelter. The flotation tails, at a pulp density of about 38 to 43% solids, together with added lime and sodium cyanide, will flow continuously through a series of eight, 9 ft diameter x 10 ft deep, agitated tanks where a simultaneous, 8 to 12-hour, carbon-in-pulp leach with 6 x 12 mesh carbon will be effected. In this leach circuit, the activated charcoal will be periodically advanced from one tank to another and retained in each leaching unit by a R.E.D. proprietory reactor design. The activated charcoal, which is expected to be loaded to about 500 oz/ton of precious metals, will be periodically removed from the first tank in the leaching system and shipped to a carbon stripping and regeneration and refining facility where pure gold and silver ingots will be produced. The leached tailings may be refloated for their remaining sulfide minerals and fine activated carbon in a second flotation step. The final tailings will be sent to an environmentally acceptable tailings pond with a proprietory R.E.D. design that allows for additional leaching and rinsing of the tailings in the tailings pond. The decanted solutions from the tailings pond, which contain small amounts of NaCN and CaO, will be collected in a holding tank and then returned to the grinding circuit.

This portable plant is to be "unitized," or built as several skid-mounted modules in a shop some distance from the mill site. The individual units that will comprise the plant described in Figure 1 would be: 1) the ore storage bin and feeder; 2) the ball mill, surge box, motor, cyclone pump, and cyclone; 3) the DSM screen, rougher and cleaner flotation cell banks, and two slurry pumps; 4) eight agitator tanks as single tanks or units of two; 5) a small bank of second stage rougher flotation cells and pump(s) (optional); 6) a power panel; and 7) a power plant (optional).

On-site preparations and facilities will include: 1) the water supply system; 2) the power system if a generator is not used; 3) trailer housing for the mill manager; 4) the plant sewer system; 5) portable office, lab, and shop buildings; 6) concrete pad and shed for mill; 7) a ramp to the ore bin; 8) the tailings pond; 9) a concrete concentrate decant pond and drying pad; 10) parking and ore storage areas; and 11) solution and pulp piping systems.

The carbon stripping and regeneration and gold and silver refining facilities will probably be located at a site other than that of the Hailey mill. It will consist principally of 1) carbon stripping columns; 2) a heated strip solution reservoir and a pump; 3) electrowinning cell(s) and rectifier(s); 4) a rotary, external-fired, carbon regeneration kiln; 5) silver and gold dissolution and precipitation equipment; 6) a crucible furnace and molds for melting and forming the gold and silver ingots; and 7) a heavy safe.

This mill design has several features that are different than most plants that have been built to process partially oxidized gold and silver ores. First, it utilizes both flotation and cyanide leaching circuits. This should ensure a maximum recovery of both the sulfide and oxide gold and silver minerals, give it a flexibility to efficiently treat a wide variety of ore types, and reduce the pulp retention time requirements in both the flotation and leaching circuits. Secondly, the flow sheet is greatly simplified; i.e., the complete or partial elimination of the crushing circuit should greatly reduce the plant and operating costs; the

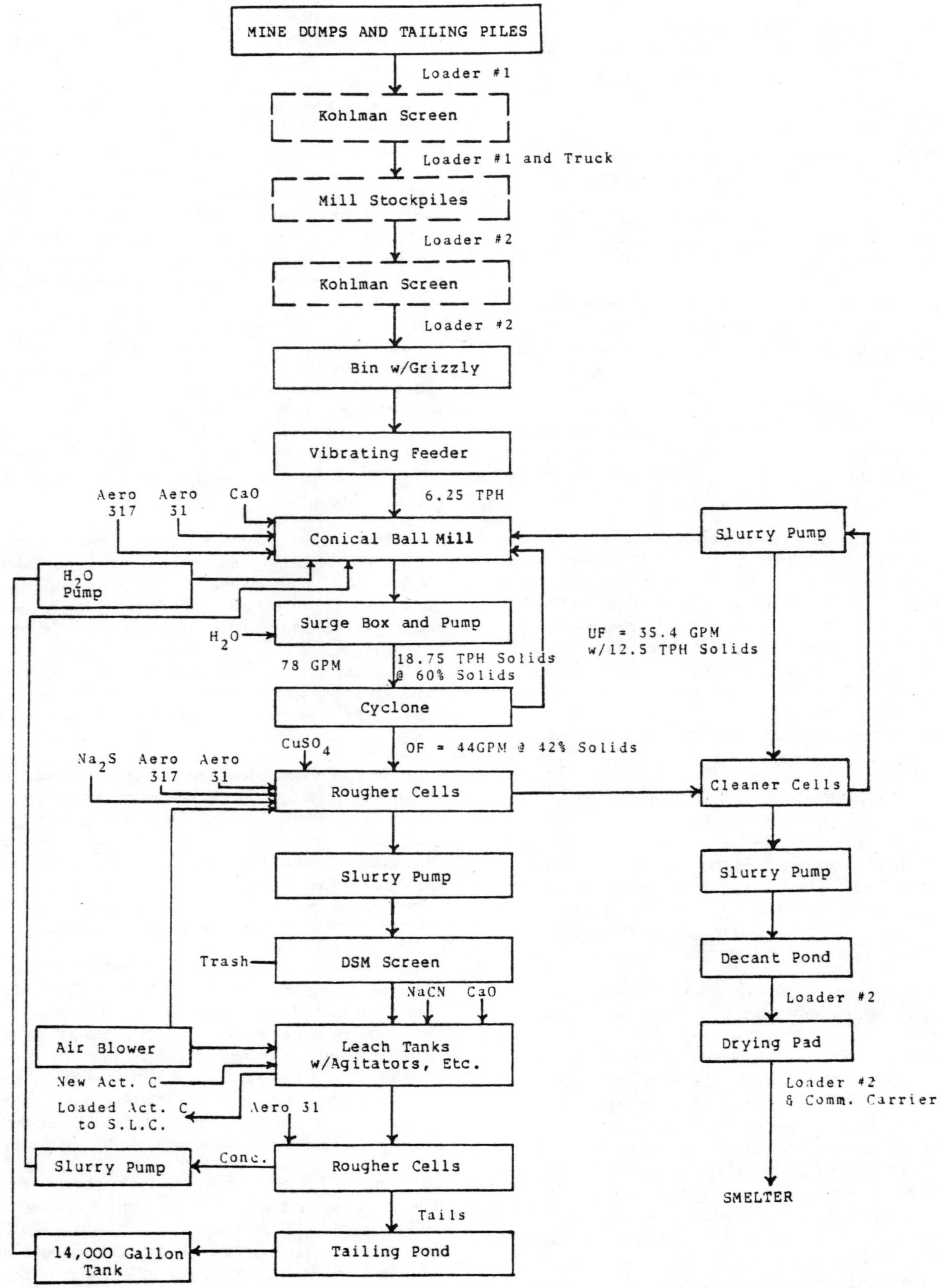

Figure 1. Flow Sheet of 150- to 225-ton-per-day Portable Mill for Processing Old Silver and Gold Mine Dumps

replacement of the flotation concentrate thickening and filtering section with a concrete decant pond and drying pad should also reduce the mill capital and operating costs and the short-term, simultaneous, carbon-in-pulp leach circuit should minimize the number and size of leach reactors that are required and reduce reagent consumptions. Thirdly, the portable nature of the plant makes it possible to build most of the plant in a shop at a greatly reduced cost, as compared with field construction; it divides its capital requirements between several sites; and keeps its resale value high.

CAPITAL AND OPERATING COSTS

Capital and operating costs have been estimated for the 150- to 225-tons-per-day Hailey, Idaho, plant and its associated carbon regeneration and stripping and gold and silver refining facilities.

The capital costs, which include used mining equipment costs (one track loader and one truck); mill equipment (new and used), modular assembly, and transportation to the millsite costs; millsite preparation costs; reagents and supplies for the mill; the carbon stripping and regeneration and refining circuit costs; engineering costs; overhead costs; environmental costs; start-up costs; and miscellaneous and contingency costs; are estimated at $1,000,000.

Mining, milling, and refining costs are estimated at $12.04/ton on old flotation mill tailings, $19.25/ton on old jig tailings, and $22.38/ton on old mine dump ores. These costs do not include truck hauling costs over four miles, royalties paid on the ores, or plant amortization costs. These ores consume about 2.7 lb of sodium cyanide per ton.

METALLURGICAL TEST WORK

Metallurgical test work has been conducted on a wide variety of oxidized gold and/or silver ore types by Resource Engineering and Development, Inc. Table 1 describes many of these tests.

Most of the flotation tests in this table were performed on ore pulps with a 40% pulp density (1,800 grams of dry ore plus 2,700 ml water), initial pH's in the 8.5 to 9 range, and with about 0.1 gram NaCN per liter of pulp. Usually, the sulfides were floated with small amounts of Aerofloat 31, Aero 317 (sodium isobutyl xanthate) and sometimes, when a bulk lead-zinc or zinc concentrate was desired, $CuSO_4 j5H_2O$ was used. Aerofroth 72 was usually used as a frother in an initial flotation step and pine oil was used in a secondary flotation step. Promoters Aero 404 and 407 were sometimes used to aid in gold recovery.

In the leaching tests, the flotation tailings that remained after sampling were placed in a four-liter beaker, NaCN, CaO, and 120 grams of 6 x 12 mesh activated carbon reagents were added, and the pulp was stirred with a mechanical agitator with air bubbling through the pulp, for 8 to 24 hours. Four-hundred-ml pulp samples were taken periodically and analyzed to determine gold and silver dissolutions, cyanide consumptions, and the degree of adsorption of precious metals by the activated carbon.

It is usually better to have the flotation step first rather than a leaching step because 1) sulfide removal reduces the cyanide consumption in the subsequent leach step, 2) the flotation concentrate is higher grade when floated before the leaching step, and 3) the amount of activated carbon required in the system is less, as much of the silver and gold is removed in the prior flotation step. In some cases, however, the cyanide leach seems to clean the mineral surfaces and thereby improves a subsequent flotation step.

The tests shown in Table 1 indicate that many oxidized silver and gold ores respond quite well to the combined flotation and cyanide leaching processes, but poorly to either process used alone.

TABLE 1
TEST WORK PERFORMED ON OXIDIZED GOLD AND/OR SILVER ORES
BY RESOURCE ENGINEERING AND DEVELOPMENT, INC.

Ore No.	Ore Source	Ore Grade, oz/ton or %		Ore Description, Grind	Process* Tested	Overall** Ag or Au Recovered, %		NaCN Consumed, lb/ton	Leach Time, hours	Overall Flotation Cleaner Concentrate Grade, oz/ton or %	
1	Hailey, Idaho	1.8	Ag	Flotation	L	67-75	Ag	2.2	8		
		1	Pb	Tailings,							
				60% -65 m***							
										15-30	Pb
2	Hailey, Idaho	3.5	Ag	Jig Tailings,	F1	42-50	Ag			50-80	Ag
		2	Pb	95% -100 m	L	78-86	Ag	2.5	8		
					F2	80-87	Ag			15-30	Pb
										50-80	Ag
										15-30	Pb
3	Hailey, Idaho	3.8	Ag	Mine Dumps,	F1	42-50	Ag			50-80	Ag
		2	Pb	95% -100 m	L	82-90	ag	2.7	8		
					F2	85-93	Ag			15-30	Pb
										50-80	Ag
										30-50	Pb
4	Central Idaho	4.0	Ag	Mine Dumps,	F1	62.6	Ag			50	Ag
		2	Pb	95% -100 m	L	82	Ag	2.5	8		
					F2	85-86	Ag			30-50	Pb
										50	Ag
5	North Idaho	0.17	Au	Stamp Mill	F	Poor					
		0.8	Ag	Tailings,	L	65	Au	5	8		
				95% -100 m		48	Ag				
										25	Pb
		3-5	Ag			26-40	Ag,			50	Ag
6	Utah	0.015	Au	Flotation	F1	Au and Pb				0.15	Au
		2	Pb	Tailings,	L	65-78		7	8		
		3	Zn	80% -100 m	F2	67-80				25	Pb
										50	Ag
										0.15	Au
		3-5	Ag			14-27	Au			low Pb	
7	Utah	2-10	Pb	Mine Dumps,	F1	and Ag			34-78	Ag	
		0.004	Au	95% -100 m	L	70-75	Ag	4	8		
		5	Zn		F2	and Au					
8	Utah	5	Ag	Underground,	F	80-83	Ag			40-80	Ag
		0.01	Au	95% -10 m							
		5	Pb	(mostly sulfides)							
						11.5	Ag				
9	Colorado	5	Ag	Mine Dumps,	F1	50	Au			21	Ag
		0.03	Au	98% -100 m	L	77	Au	8	24		
						50-62	Ag			21	Ag
					F2	50-62	Ag				
10	Southern California	4.5	Ag	Mine Dumps,	F	None					
				95% -100 m	L	Poor	8	24			
11	Southwest Montana	30.3	Ag	Mined Surface	F	60	Ag			565	Ag
		0.02	Au	Ore,	L	87	Ag	10-15	24		
		0.54	Cu	80% -100 m							

TABLE 1 (Continued)

Ore No.	Ore Source	Ore Ore Grade, oz/ton or %	Overall** Description Grind	Ag or Au Process* Tested	NaCN Recovered, %	Leach Consumed, lb/ton	Overall Time, hours	Flotation Cleaner Concentrate Grade, oz/ton or %
12	Central Utah	0.13 Au 0.6 Ag	Mined Surface, Jasperiod, 99% -100 m	L	88 Au 40 Ag	6	24	
13	Southern Nevada	0.03- 0.07 Au 0.6 Ag	CN Leach Tailings, 90% -100 m (no regrind)	L	60-80 Au 20 Ag	4	12	
14	Nevada	1.81 Ag 0.062 Au	CN Leach Tailings, 98% -100 m	F L F	Poor 58 Ag 92 Au	6.5	8	
15	Canada	0.10 Au	Tailings, reground to 98% -100 m	F L	Poor 74.7 Au 43.5 Au w/o regrind	3	16	
16	Utah	1.5 Ag	Flotation Tailings, 90% -100 m (no regrind)	F L	Poor 35 Ag	6	24	
17	Idaho	1.5 Ag 0.018 Au 0.5 Pb 0.1 Cu 0.6 Zn	Dump Material, 95% -10 m	F	70 Ag 50 Au 81 Pb 32 Cu 11 Zn			1 Au 69 Ag 42 Pb 3.2 Cu 6.3 Zn
18	Southern California	0.03 Au 0.2 Ag	Placer Slimes, -100 m	L	95 Au 95 Ag	2	8	

Chapter 8

THE CARBON–IN–PULP PLANT OF THE MASBATE GOLD OPERATIONS PHILIPPINES

R. S. Pizarro
F. V. Antonio
Atlas Consolidated Mining and Development Corp.
Manila, Philippines

ABSTRACT

Masbate Gold Mines, dormant since WWII, are now back in operation. The ore treatment plant incorporates carbon–in–pulp processing, utilized because of the ready availability of coconut shell charcoal in the Philippines and the high clay content of the ore. A detailed description of the plant, including the surprisingly fast and efficient alcohol stripping facility is presented. Operating data for the plant is provided together with an interesting performance evaluation of mechanical versus air-lift agitated leach tanks.

Introduction

The Masbate Province, Philippines, has been a gold producer for many years. Several gold properties were profitably operating until the outbreak of World War II in 1941. One of these properties was the Masbate Consolidated Mining Company that at the time showed an impressive production rate of more than 4,000 metric tons of ore per day and one of the lowest milling costs on record in the country at that time.

Interesting enough, it was the milling equipment left abandoned at Masbate that gave birth to one of the giant copper producers in ASIA. In 1955, Atlas Consolidated Mining and Development Corporation started a new operation at its copper property in Cebu with the seven 2.4m Ø x 3m long (8'Ø x 10' long) ball mills left abandoned at the old Masbate Gold Mine. Now, 25 years later, Atlas is processing over 100,000 metric tons of copper ore per day through its three copper concentrators, and, the seven old mills are still grinding, copper ores now.

The recent dramatic increase in the price of gold compelled Atlas Management to look back to the dormant gold fields of Masbate and in a record time a decision was made:

REACTIVATE THE MASBATE GOLD MINES !

The decision was made at the end of 1978. Fourteen months later, on January 31, 1980, the first bullion bar was poured at the new Masbate Gold Operations.

A reborn mine is again giving life to a forgotten, but beautiful island in the Philippines.

Geology

Four distinct rock types have been identified in the area, namely: tuffaceous agglomerates, andesitic agglomerates, metasedimentary rocks and hornblende augite andesite porphyry.

The tuffaceous agglomerate occupies the northern and western flanks of the mine area while the andesitic agglomerates are localized distinctly in the central portion of the volcanic plug within the mine site. The eastern boundary of the area is occupied by metasedimentary rocks whereas the hornblende augite andesite porphyry occurs as intrusive within the andesite agglomerate and tuffaceous agglomerate.

Faults and vein walls are the controlling factor in ore trends as the gold and silver are more concentrated within the vein fault system and not so much in the wall rocks.

The main shear in the mine area is the NNW trending faults which was truncated by a WSW trending fault.

Hydrothermal alteration zonning is defined by the vein wall contacts. A narrow zone of overlapping argillization and silicification at or near vein contacts grades to chloritization and silicification away from the contact of vein. This further grades into chlorite-epidote alteration as we move further away from the vein. The major alteration products usually associated with ore values in the wall rocks are argillization and silicification.

Gold and silver are microscopically disseminated (grain size 10-20 microns) within the vein and immediate wall rock together with pyrite. High grade ores are usually hosted in the brecciated and cellular variety of quartz vein with abundant Manganese Oxide.

At present, the Masbate orebody is being mined by open pit methods.

The ore body is estimated to contain 8 million tons of minable ore with an average gold assay of 2.4 g/t and 3.4g/t of silver (0.08 oz/tonne Au and 0.10 oz/tonne Ag respectively.

The Mine

The pit area covers an area of approximately 18 hectares.

The gold bearing orebody is found in an area overlain by tuffaceous agglomerate on its northern and western flanks while the eastern flank is surrounded by metasedimentary rocks.

The pit has been designed under the following specifications:

1. Pit Slopes = 45 degrees
2. Bench = 70-75 degrees
3. Final Berm Width = 6-7 meters
4. Bench Heights = 10 meters
5. Maximum grade of ramps = 8%
6. Minimum Width of ramps = 18 meters

Preproduction stripping of the main Pit started in October, 1979 and was completed early in January, 1980. Some 830,000 metric tons of material were stripped.

Percent dilution, in terms of internal waste is estimated at 7 percent of tonnage.

The average stripping ratio will be 1.1:1.

Drilling and Blasting Operations

Drilling requirements are provided by one Rotamec 1702 drill and one CP 750 Reich drill as back up. Both are capable of producing 17.1 cm. diameter holes (6-3/4 in.).

Drill holes for blasting are spaced at 4.0m x 5.0m. Drilling depth including subgrade is 12 m.

Two smaller drills capable of producing 7.62 cm. diameter holes (3 in. holes) are used in secondary drilling and other related projects. The powder factor ranges between 4-6 metric tons of ore per kg and 6-10 metric ton of waste per kg.

Loading and Hauling

Blasted pit materials are loaded into three -75 Wabco rear dump trucks utilizing one 4.58 cu.m. (6 cu. yd.) B.E. electric shovel. This shovel has a rated capacity of 10,000 metric tons per day, although some 8,000 metric tons will have to be moved to satisfy the stripping ratio.

Average truck capacity is rated at 3,000 metric tons per day; Mill operation is rated at 3,500 metric tons per day.

Grade Control Operations

Blast hole cuttings are sampled, split and reduced to some two kg and sent to the Assay

Laboratory.

Daily advances of the shovel are directed by the Assay Laboratory results. All materials are assayed before leaving the pit. According to assay results the material is either sent to the Mill or to waste.

Short range mining plans are prepared monthly for three consecutive months. Long range planning is spaced quarterly for the first year and semi-annually thereafter for the next five years.

The Mill

A conventional Three Stage Crushing Plant has been adopted for the Masbate Gold ore. (Please see Figure No. 1)

A 0.91 m x 1.22 m Traylor Jaw Crusher reduces the ore to a minus 12.7 cm.(5 in.) product. The ore is conveyed to a Washing Plant consisting of two 2.44 m x 3.66 m (8'x12') scubber mills. These facilities were included in the flow-sheet in anticipation of high clay content in certain areas of the mine. The Washing Plant can also be by-passed as needed.

So far, Masbate has operated most of the time excluding the Washing Plant experiencing minor ore handling problems in the Crushing Plant.

A coarse ore stockpile, 12,000 t capacity, regulates the ore flow to the secondary Crushers (two -1.67 m (5-1/2 ft) Symons Standard) and to the Tertiary Short Head Cone Crushers (two 1.67 m (5-1/2 ft) Symons Crushers).

Final product of the crushing plant is minus 9.5 mm (3/8 in) material. The crushed ore is conveyed to two enclosed F.O.B. with approximately 2,000 metric ton live capacity each.

The ore is advanced to the grinding plant consisting of four Hardinge Ball Mills 3.2 mØ x 4.87 m (10-1/2' Ø x 16') equipped with 671kw motors. The Bond Index for the ore is variable and ranges from 15 to a high 20 kWh/ton. The classification is done on 50.8 cm (20 in.) cyclones. Cyclone overflow at about 24% solids yields the required mesh of grind established at 80% passing 200 mesh. Coarser grind invariably results in higher residue assays in the final tailings. Only lime is added to the ball mill feed to maintain a pH of 10.5.

Sodium Cyanide in solution form is added ahead of the leaching circuit. Sodium Cyanide consumption proved to be higher when added to the ball mills without noticeable improvement in gold dissolution.

Table No. 1 shows the main reagents average consumption at the Masbate Gold Operations to date.

The cyclone overflow product is directly sent by gravity to a Primary Thickener where the pulp density is raised to 45% solids. Thickener overflow is reused as ball mill grinding water.

Sodium Cyanide solution at a rate of 0.9kg/t of ore (2 pounds/tonne) is added to the thickener underflow pump ahead of the Leaching Circuit.

This circuit provides a 24-hour retention time and consists of eleven 9.14 m Ø 9.14 m (30' Ø 30') agitator tanks. Air is sparged inside the tanks to provide the necessary oxygen for dissolution of the gold and silver values. Agitation is imparted by mechanical, axial flow, rubber lined propellers.

Gold dissolution in this circuit averages 88.7 while silver dissolution averages 66.0 percent.

The leached slurry is now advanced by gravity to the Carbon-in-Pulp Circuit.

Carbon-in-Pulp Circuit

The slurry at a rate of approximately 3.785 cu. m. per minute (1,000 GPM) is placed in contact with activated carbon. A simplified flow-sheet of the CIP Plant is shown in Figure 2.

For years it has been known of the affinity of activated carbon to adsorb preferentially the gold and silver in cyanide complex form. This operation takes place in five circular tanks that provide a total pulp retention time of 5.5 hours. The carbon advances countercurrently to the pulp. The carbon is maintained in suspension in the pulp by an internal air-lift and mechanical agitation of the pulp, either with rakes or with axial flow propellers. This last arrangement is preferred because in case of power outages the pulp can be easily repulped after the power is restored. By means of an external air-lift, the pulp is elevated to the top of two back-to-back DSM Screens, such that the granular activated carbon (-6/ 10) mesh is retained on the screen surface and the finely ground pulp is piped down to the next tank in the circuit. Both, carbon and pulp, advance countercurrently by gravity. Carbon is advanced such as to satisfy daily production requirements. An adequate amount of fresh or reactivated carbon is added daily to the last tank in the circuit.

Carbon concentration in the pulp averages 24 grams of carbon per liter of pulp. Carbon loading,expressed in grams per metric ton of carbon is shown on Table 2. Typical solution gradient curves are shown on Figures 3 and 3-A. After carbon reaches a satisfactory loading, it is washed on a vibrating screen and then educted to the Carbon Desorption Circuit. A typical two-ton batch is stripped in about 14 hours. However, present facilities are such that six tonnes of carbon can be stripped in one day if necessary.

Typical stripping operation parameters are shown in Table No. 3. An alcoholic solution containing ethanol, caustic soda and sodium cyanide is heated to 90°C and pumped through the Carbon Bed. The solution has been heated with steam in a stainless steel tank. The steam is externally injected through a steam jacket of the "clamp-on" type.

For best efficiency, desorption columns, alcohol tank and piping throughout have been thermally insulated.

Adequate and basic instrumentation has been provided for the efficient operation of the desorption process.

The most critical factor for an efficient desorption rate is the temperature. A drop in temperature by 10°C may increase the time for desorption by 6 hours (from a typical 14 hours to 22 or 24 hours). Figure No.4 shows a typical stripping solution curve.

Flowrate is also critical and is carefully

monitored during the operation.

Floor washing and excess water generated during carbon handling is recirculated within the plant. Carbon is handled in slurry form using water jet eductors. These operate very efficiently. A booster pump rated at .38 cu. meters per min. (100 GPM) and 10 bars (145 psi) provides the necessary water flow and pressure for handling the carbon throughout the plant.

Electrowinning Circuit

The carbon stripping process requires the continuous removal of gold and silver by passing the pregnant solution through a battery of electrolytic cells.

The process continues until the gold and silver in the solution is dropped below five (5) grams of gold per metric ton of solution (5 ppm).

Gold and silver are deposited on steel wool that serves the purpose of cathodes. Anodes are made of stainless steel screen mesh .

Three rectangular cells have been designed to provide 6 min. retention time per cell. The cells are made of stainless steel insulated internally with FRP.

Direct current to the cells is provided by a rectifier rated for an output of 300 amp and 12 volts per cell.

There are two banks of three cells each. Each bank is alternatively used every month.

This method provides a very convenient system for accurately accounting metals production for any particular month. Present practice consists of removing cathodes three times a week for a total production of approximately 186.6 kg (6,000 ounces), of gold and 124.4 kg (4000 ounces) of silver per month.

The efficiency of this circuit is very high and averages approximately 97% for gold and 90% for silver.

The fire hazard involved in handling alcoholic solutions demands special precautions to avoid overheating in localized areas, rapid removal of alcohol vapors, adequate provision of firefighting equipment and close supervision of the operations around the clock. So far, no special problems have been experienced in this area.

Carbon Reactivation

This is the most important aspect for a successful CIP Plant. No matter how good the new, fresh carbon is, if the carbon is not properly reactivated, the efficiency of the carbon adsorption circuit will be low and consequently, high gold and silver losses will be experienced in the CIP Plant.

Our Rotary Kiln provide 30-minute retention time ; reactivation temperature is 900°C. The Kiln 0.91 m Ø x 6 m long (3' Ø x 20' long) is externally heated with three burners equally spaced along the kiln.

Temperature controllers and indicators are installed to maintain the required operating conditions. The kiln is equipped with a variable speed screw feeder to regulate carbon feeding to the kiln.

Every batch of stripped carbon is first water drained, then water rinsed, then steam scrubbed and finally educted to the Rotary Kiln.

The quality of the reactivated carbon is periodically checked and normally compared with fresh activated carbon.

Locally manufactured activated carbon is used in the Masbate Operations with satisfactory results. Local carbon is produced from coconut shells which is abundant in the Philippines.

Fine carbon losses are daily measured and carefully evaluated. Although coconut is softer than peach pit carbon or walnut shell carbon, our consumption averages 0.10 kg per ton of ore, which seems to be somewhat higher than similar operations elsewhere.

Typical results on activated and reactivated carbon are shown in Table No. 4.

The regenerated carbon discharges by gravity into a conical tank where it is water cooled ; then ejected to a washing screen before being reintroduced to the Carbon Adsorption Circuit.

Refinery

The loaded cathodes are melted in one vertical crucible type furnace (size 150). The furnace uses fuel oil No. 6.

The most important impurity to eliminate is iron originating from the steel wool. The iron is fully oxidized with sodium nitrate,borax and silica, and transformed into iron oxide for removal in the slag.

The liquid mass is poured into a slag pot (one cubic foot capacity). The gold and silver metals accumulate at the bottom of the slag pot. After cooling, a bullion button is obtained.

This bullion still contains minor amount of impurities so the buttons are re-melted in a separate vertical furnace where, by the addition of borax, silica and niter, the remaining impurities are removed.

The bullion is poured into a 31.1 kg (1000 ounce) bar mold. Average bullion fineness is 660 for gold and 330 for silver.

The bars are delivered to the Central Bank Refinery of the Philippines.

General Comments

The Masbate Carbon-in-Pulp process plant has passed through the usual problems of starting-up a new plant, and although all bottlenecks have not been finally adjusted, operations have reached a fairly normal level of production.

Table No. 5 shows typical operations parameters of the Masbate Gold Operations.

A comparison between mechanically agitated vs. rake agitated-central air-lift tanks is presented in Table No. 6.

One point worth of mentioning here is that gold dissolution continues in the Carbon Adsorption Circuit.

In Masbate Gold Operation, the gold dissolution in the Carbon Adsorption Circuit increases by about 2% as measured by a reduction in the residue assay of the CIP tails. This is attributed to the strong aeration imparted by the air-lifts and the extended retention time (5.5 hours in this case).

The use of alcohol (ethanol) in the stripping of gold has proved to be practical, simple and efficient. As it can be observed in Table

No. 3, the carbon tails assay average 44.5 g/t and 112.9 g/t for gold and silver respectively. These are highly acceptable values taking into consideration that the gold and silver values are returned to the circuit.

The average stripping time per 2 ton batch is 14 hours, which campares very favorably with the 50 hours required by the Zadra process.

We have been experimenting with steam and hot water in an effort to decrease even more the stripping time.

Certainly, the mechanism of gold adsorption and desorption from granular activated carbon remains rather speculative at the present time.

However, results obtained in industrial scale demonstrate that activated carbon offers a practical, simple and efficient alternative for the recovery of precious metals over the conventional zinc dust method.

Acknowledgement :

The authors wish to express their sincere thanks to the Management of Atlas Consolidated Mining and Development Corporation for permission to publish this article.

Appreciation is also extended to the Metallurgical and Engineering Staff of ACMDC for their invaluable help during the development, design and construction of the Masbate Gold Operations.

TABLE NO. I

MGO AVERAGE REAGENT CONSUMPTION

	Reagent	Consumption	
1.	Sodium Cyanide	0.89 kgr/mt	(ore)
2.	* Lime	4.88 kgr/mt	"
3.	Activated Carbon	0.10 kgr/mt	"
4.	Ethanol	0.40 l/mt	"
5.	Caustic Soda	0.04 kgr/mt	"
6.	Boráx	0.01 kgr/mt	"
7.	Silica	0.007kgr/mt	"
8.	Sodium Nitrate	0.01 kgr/mt	"
9.	Steel Balls	1.5 kgr/mt	"

*Lime contains 60% available CaO.

TABLE NO. 2

CARBON LOADING - SOLUTION GRADIENT DATA

With Four Tanks	Carbon Loading		Solution Gradient	
	Au (g/t)	Ag (g/t)	Au (g/m^3)	Ag (g/m^3)
1	1950	1471	0.93	1.56
2	1078	958	0.47	1.32
3.	486	479	0.27	1.10
4.	205	274	0.08	1.09

Operational Data for June, 1980

Feed Solution:
Au 2.02 g/m^3
Ag 3.15 g/m^3

Carbon Concentration: 25 g/l
Adsorption Recovery:
Au 96 %
Ag 65.4%

With Five Tanks	Carbon Loading		Solution Gardient	
	Au (g/t)	Ag (g/t)	Au (g/m^3)	Ag (g/m^3)
1	1710	1300	0.52	1.46
2	684	718	0.24	1.19
3	342	480	0.15	1.10
4	171	307	0.09	1.02
5	137	205	0.05	0.94

Operation Data for September, 1980:

Feed Solution:
Au: 1.89 g/m^3
Ag 2.83 g/m^3

Carbon Concentration: 24 g/l
Adsorption Recovery:
Au: 97.3%
Ag: 66.8%

TABLE NO. 3

TYPICAL STRIPPING OPERATION DATA

Carbon Stripped per batch	2.0 t
Carbon Head Assay:	
Gold	1,710 g/t
Silver	1,300 g/t
Carbon Tail Assay:	
Gold	44.5 g/t
Silver	112.9 g/t
Recovery:	
Gold	97.4%
Silver	90.7%
Time to complete (average)	14 hours
Temperature	90°C

STRIPPING TIME, HRS.	PREGNANT SOLUTION (Au. g/m^3)
0.5	20
2.0	250
4.0	130
6.0	70
8.0	38
10.0	18
12.0	8
14.0	4

TABLE NO. 4

CARBON REACTIVATION DATA

Kiln Capacity	4 tpd
Kiln Size (s.s. shell)	.91 m ø x 6m long
Retention Time	30 min.
Temperature	900°C
Gold Adsorption Capacity of Reactivated Carbon	88.0%
Gold Adsorption Capacity of Fresh Carbon	91.0%
Gold Adsorption Capacity of Un-reactivated Carbon	68.0%
Test Conditions:	
Pulp Volume	1 liter
Pulp Density	1370 gpl
Carbon Concentration	10 gpl
Agitation Time	1 hour

TABLE NO. 5

MGO TYPICAL OPERATING PARAMETERS

1. Mill Design Capacity	3500 tpd
2. Head Assays:	
Au	2.60 g/t
Ag	3.60 g/t
3. Ball Mill Feed:	
Size: % - 1.27 cm	2.0%
W. I.	15.15 Kwh/t
4. Cyclone Overflow:	
Percent Solids	24%
Size - % -74 microns	80 %
5. Leaching Circuit	
Feed % Solids	45 %
Residence Time	24 hrs.
Leach Residue Assays:	
Au	0.29 g/t
Ag	1.22 g/t
% Dissolution:	
Au	88.8%
Ag	66.1%.
6 Carbon Adsorption Circuit	
Solution Feed Assay:	
Au	1.89 g/m^3
Ag	2.83 g/m^3
Solution Tails Assay:	
Au	0.05 g/m^3
Ag	0.94 g/m^3
Percent Solution Recovery:	
Au	97.3 %
Ag	66.8%

NOTE: Data for September, 1980

TABLE NO. 6

COMPARISON BETWEEN MECHANICAL AGITATORS VERSUS RAKE TYPE AGITATORS

Factor	Mechanical Agitator	Rake type Agitator w/ Center Air Lift	Comments
1. Dependability	Exceilent	Poor	In case of total power outages, the pulp settles very fast. With the rake rake type agitators it takes several hours to return the tanks to operation. In case of the Mechanical Agitators, operations are normalized as soon as power is restored.
2. Operating Efficiency	Excellent	'atisfactory	Build up of carbon in the bottom of the tank is minimal with the mechanically agitated tanks; on the other hand, with the rake type agitator, a layer of about 6-8 inches of gold loaded carbon is normally formed under the rakes.
3. Mechanical Efficiency	Excellent	Satisfactory	With the rubber lined impeller of the Mechanical agitator wear is minimal with the rake type agitators, rakes wear out fast and supporting chains or cables must be periodically inspected and replaced.
4. Power Consumption	High	Very Low	Our mechanical agitators use 60 HP motors while the rake type agitators use only a 5.5 HP motor. We also need to add the horsepower required to generate some 60 CFM of low pressure compressed air required by the rake type agitator.
5. Metallurgical	Excellent	Satisfactory	Recovery: Au Ag Mechanical 72.5% 48.4% Rake Type 54.0% 50.5%

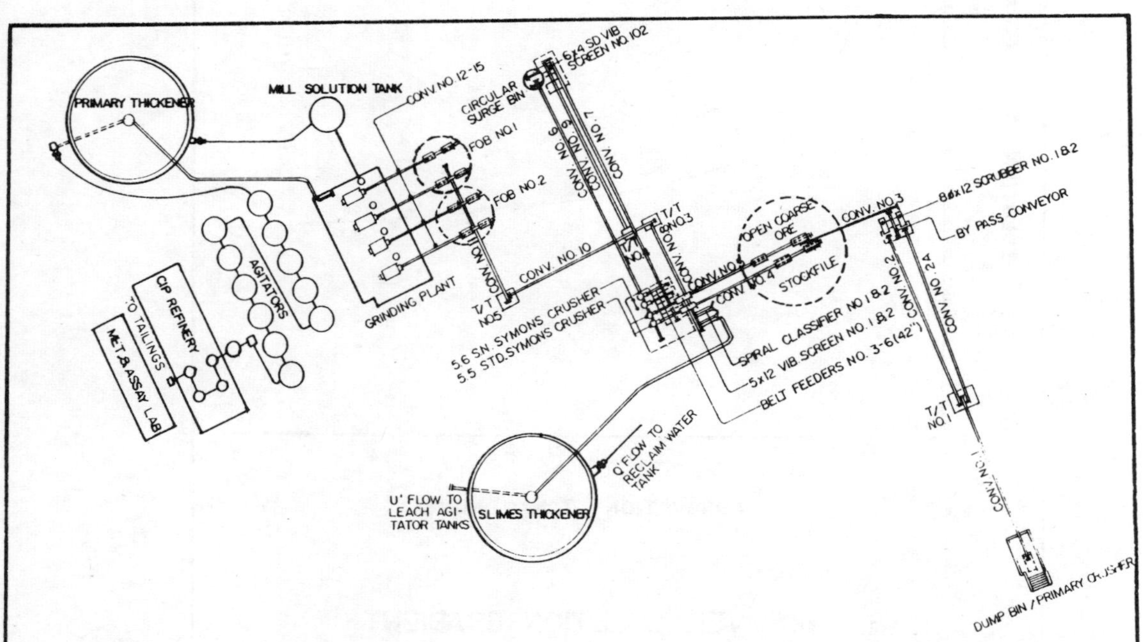

FIGURE NO. 1
MASBATE GOLD OPERATION
GENERAL PLANT LAYOUT

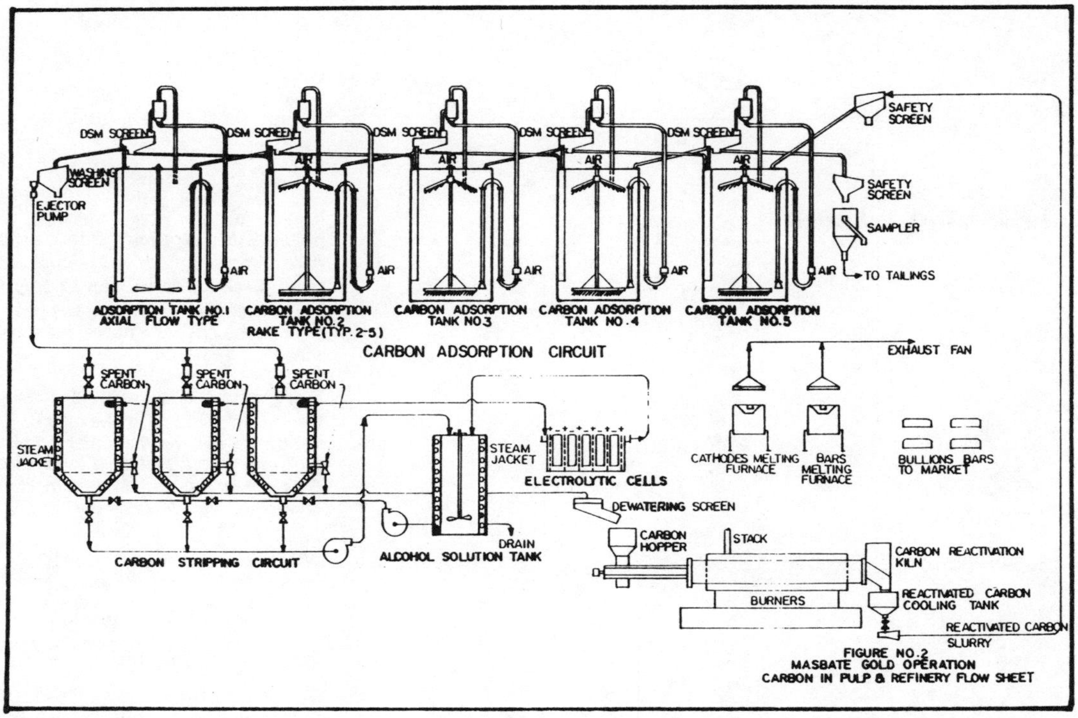

FIGURE NO.2
MASBATE GOLD OPERATION
CARBON IN PULP & REFINERY FLOW SHEET

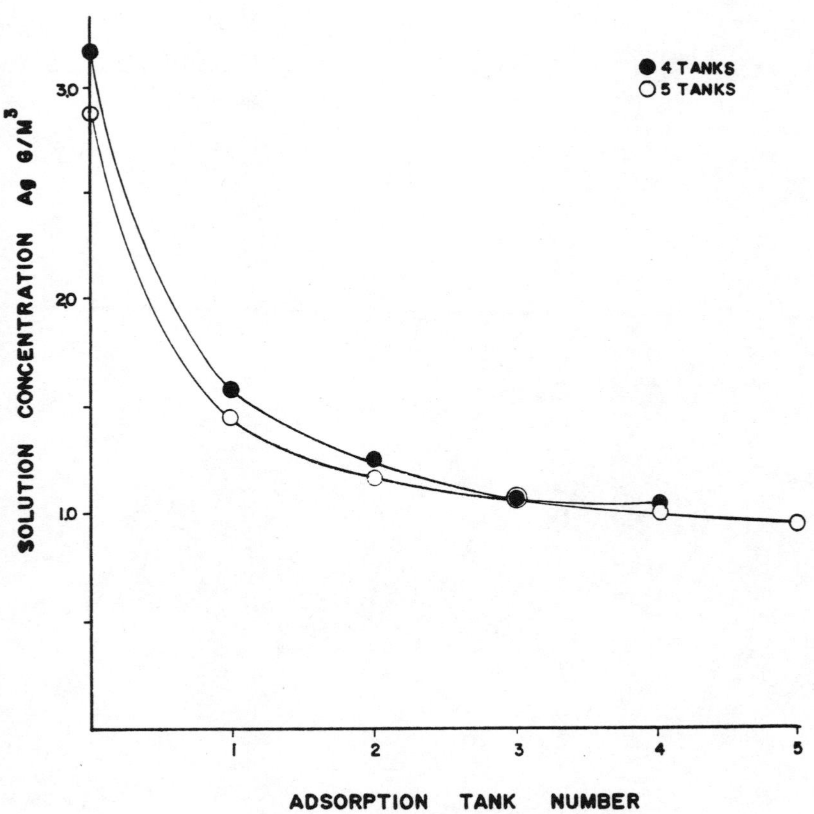

FIG. No.3A SILVER SOLUTION GRADIENT

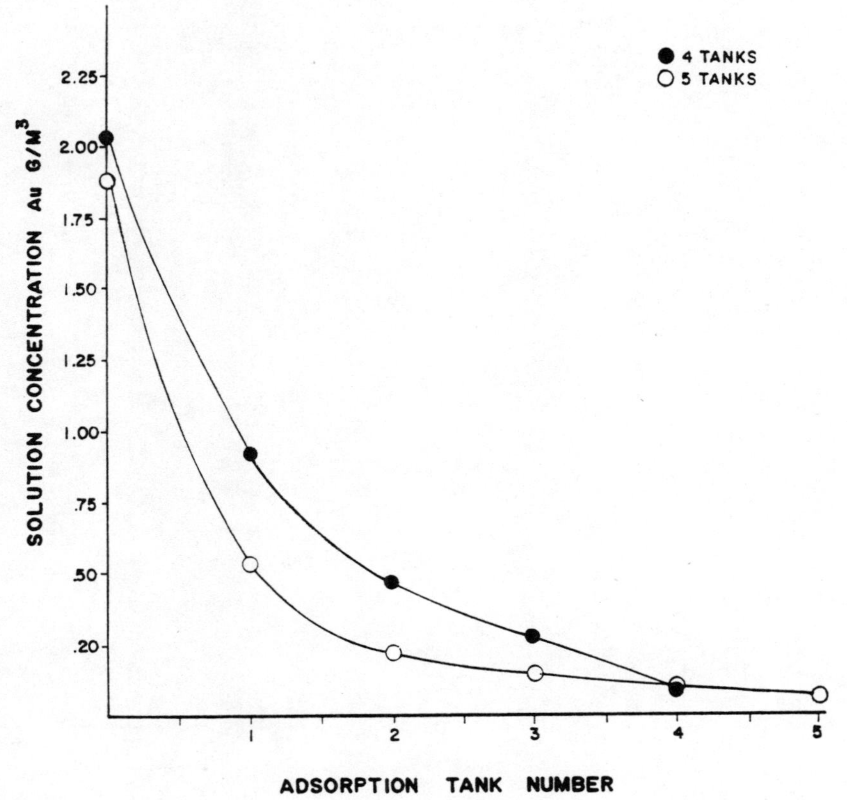

FIGURE No.3 GOLD SOLUTION GRADIENT

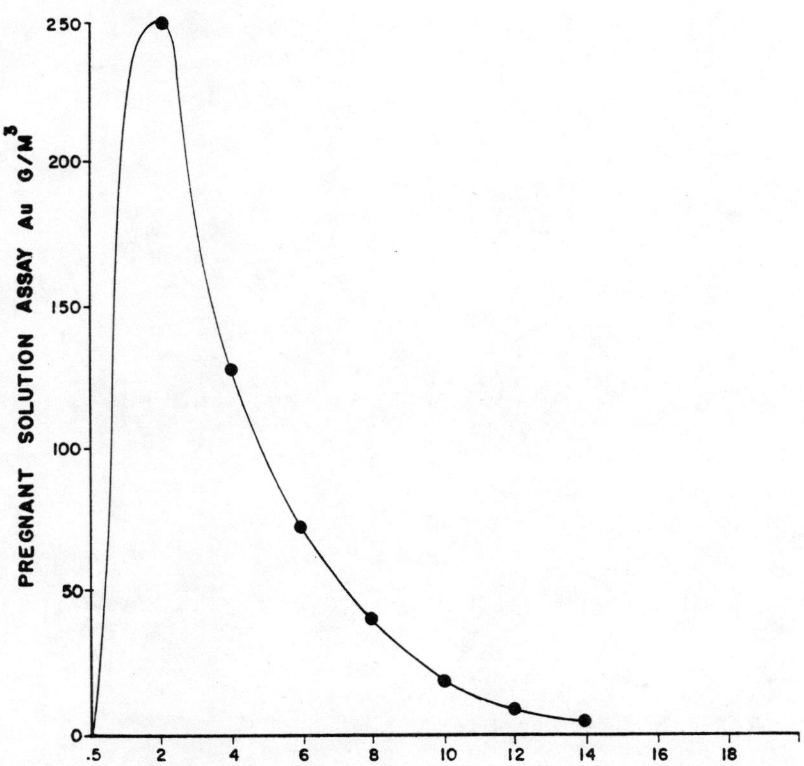

FIG. No.4 TYPICAL STRIPPING SOLUTION CURVE

Chapter 9

DESTRUCTION OF CYANIDE IN GOLD AND SILVER MINE PROCESS WATER

O. B. Mathre
F. W. DeVries
E. I. du Pont de Nemours & Co.
Wilmington, DE 19898

ABSTRACT

Addition of Kastone® peroxygen compound and formaldehyde to cyanide gold extraction process effluents, containing at least 5 mg soluble copper per liter, will oxidize acid-dissociable cyanide to levels less than 0.2 mg per liter, without appreciable oxidation of thiocyanate. Most of the copper precipitates and is removed by filtration of the treated solutions. The treatment can be completed in less than 2 hours when the excess of hydrogen peroxide over the cyanide is about 75–100 mg per liter. The excess hydrogen peroxide destroys residual formaldehyde and then is catalytically decomposed by the copper-containing solids that separate, so it will not interfere with subsequent operation.

A pilot plant using this technology has been operated for more than four months at rates from 1–2 gpm. Clear, filtered effluent containing less than 0.2 mg acid-dissociable cyanide per liter has been produced.

Introduction

The extraction of gold and silver from ore with alkaline cyanide solutions has been an important process for about ninety years. The waste water from this operation generally contains variable amounts of sodium cyanide, acid-dissociable complex metal cyanides and thiocyanate. Some waste waters may also contain oxidizable sulfur species such as sulfide, polysulfide, sulfite and thiosulfate. Routes for the formation of copper cyanide, refractory iron cyanides and thiocyanates in gold and silver extraction are shown in figure 1.

Figure 1

Gold and Silver Extraction

$$FeS + xs\ CN^- + Air \xrightarrow{OH^-} Fe(CN)_6^{-4} + CNS^-$$

$$Copper\ (Carbonate\ ore) + xs\ CN^- \xrightarrow{OH^-} Cu(CN)_2^- + CNO^-$$

$$S_{n+1}^{-2} + n\,CN^- \longrightarrow n\,CNS^- + S^{-2}$$

State and Federal guidelines and regulations propose or specify limits on the concentration of cyanide, free and complexed, that may be discharged to streams and in some cases sanitary sewers. The limits set by the regulatory authorities are generally at levels that require some type of cyanide analysis methodology that is used to define the cyanide content of the stream.

Oxidation with Chlorine

Alkaline chlorination is commonly used to destroy cyanide in waste process water. It destroys free cyanide and cyanide complexed with common metals such as cadmium, chromium, copper, gold, nickel, silver and zinc. It does not destroy the cyanide present as ferrocyanide or ferricyanide. It also reacts with thiocyanates and other components of waste streams, which may not require destruction and can produce undesirable by products. The net chemical reactions are shown in figure 2.

Figure 2

Process Effluent Treatment Alkaline Chlorination

$$CNS^- + 4\,OCl^- + 2\,OH^- \longrightarrow CNO^- + SO_4^{-2} + 4Cl^- + H_2O$$

$$CN^- + OCl^- \longrightarrow CNO^- + Cl^-$$

$$2\,Fe(CN)_6^{-4} + OCl^- + H_2O \longrightarrow 2Fe(CN)_6^{-3} + 2OH^- + Cl^-$$

$$2\,Cu[I] + OCl^- + H_2O + 2OH^- \longrightarrow 2Cu(OH)_2 + Cl^-$$

It was known that thiocyanate, a common by-product of cyanide ore leaching, would react with hypochlorite under conditions used for alkaline chlorination of cyanide compounds. Since waste water from cyanide gold and silver extraction processes usually contains thiocyanate in substantial quantities, it is necessary to add enough alkaline hypochlorite to destroy both the compounds. These reactions proceed in steps, but we will consider the net reactions shown in the first two equations in figure 2. It is apparent that the presence of moderate amounts of thiocyanate in a waste stream will substantially increase the chlorine demand of the waste stream.

Both synthetic and real process solutions were subjected to standard alkaline chlorination conditions, except that the amount of hypochlorite added was limited to that required for cyanide destruction plus an excess that ranged from 10-100% of the calculated value. Analysis of the solutions for cyanide and thiocyanate showed that the thiocyanate reacted preferentially with the hypochlorite and that the cyanide concentration was almost unchanged. This observation confirms other data that enough hypochlorite must be added to the waste stream to destroy both the thiocyanate and the cyanide to reduce the measured levels of acid-dissociable cyanide in the treated effluent. Figures 3 and 4 show typical results.

Figure 3

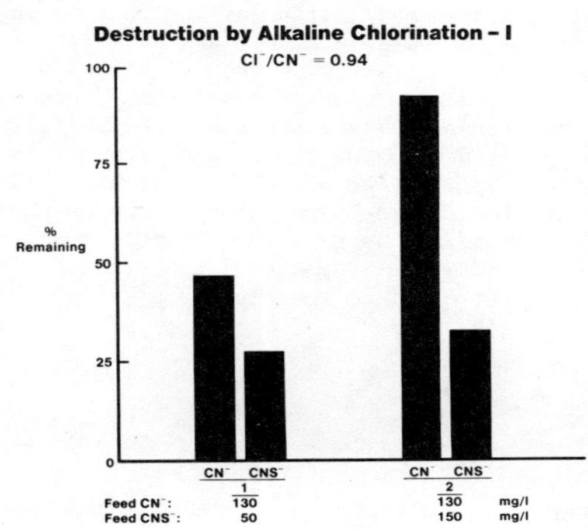

Figure 4

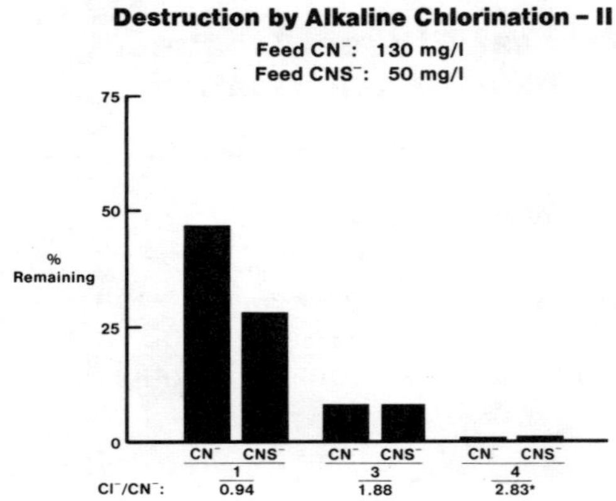

Figure 5

Process Effluent Treatment - Hydrogen Peroxide

$$2\,Cu(CN)_2^- + 5H_2O_2 \longrightarrow 2Cu^{+2} + 4CNO^- + 5H_2O$$

$$2Cu^{+2} + 5CN^- + 2OH^- \longrightarrow 2Cu(CN)_2^- + CNO^- + H_2O$$

$$CN^- + H_2O_2 \longrightarrow CNO^- + H_2O$$

$$CN^- + HCHO + H_2O \longrightarrow HOCH_2CN + OH^-$$

$$HOCH_2CN + H_2O \xrightarrow[OH^-]{H_2O_2} HOCH_2CONH_2$$

Even if sufficient chlorine has been added to destroy the thiocyanate as well as the cyanide, there is a further problem: It has been found in analytical method evaluation that solutions containing 0.1-5 mg cyanide per liter and more than 10 mg thiocyanate per liter will show a positive test for cyanide after treatment with an excess of chlorine even for 30 minutes. The analytical samples were dechlorinated with either ascorbic acid or sodium arsenite, preserved with sodium hydroxide and analyzed for acid-dissociable cyanide and for total cyanide by the official methods. Cyanide values greater than 0.1 mg/l were obtained. This suggests that alkaline chlorination of waste waters that contain substantial quantities of thiocyanate can yield effluents that may not meet regulatory requirements for discharge unless different alkaline chlorination process conditions or sample-handling techniques are used.

Oxidation with Peroxide

It was reported by Masson in 1907 that hydrogen peroxide would react with alkaline sodium cyanide solutions.(1) The kinetics of the reaction given in the paper showed that the time required for reduction of cyanide from 50 mg/l to less than 0.5 mg/l would be more than 8 hours even when the initial molar ratio of hydrogen peroxide to cyanide was as high as five. During the period 1968-1971 it was reported that the reaction between hydrogen peroxide and cyanide was promoted by addition of compounds such as formaldehyde and soluble copper salts.(2-5) The reaction rate was increased so that it became a viable candidate for destruction of cyanide in process effluents. It also was shown that substantially improved hydrogen peroxide reaction efficiency could be obtained by addition of decomposition stabilizers. This stabilized hydrogen peroxide, known as Kastone® peroxygen compound, is used in combination with formaldehyde promoter for commercial destruction of cyanide. The net chemical reactions are given in figure 5.

Since Kastone® technology was developed for the plating industry, the effect of thiocyanate on the process had not been examined. We found that cyanide was selectively oxidized in alkaline solution by formaldehyde or soluble-copper promoted hydrogen peroxide in solutions containing thiocyanate. Analysis showed that generally less than 10 percent of the thiocyanate was destroyed. The cyanide found in the treated solutions was less than 0.2 mg per liter. The copper-promoted system generally gave cyanide residual values below 0.05 mg per liter. The results obtained for several typical tests are given in figure 6.

Figure 6

Lab Tests - Synthetic

	mg/l			
	CN^-	CNS^-	H_2O_2	Cu
Feed	20	50	155	10
Effluent @ 36 min	0.04	49	-	-
Feed	10	100	155	5
Effluent @ 29 min	0.01	99	-	-
Feed	20	50	211	10
Effluent @ 211 min	0.03	48	-	-
Feed	20	50	196	2
Effluent @ 142 min	≤0.01	51	-	-
Feed	10	50	128	5
Effluent @ 57 min	0.01	50	-	-

Treatment of Field Samples

Cyanide gold and silver mine process effluents frequently contain soluble copper cyanide complexes. Synthetic process solutions containing sodium cyanide, copper cyanide complex, ferrocyanide and thiocyanate were treated with Kastone® peroxygen compound at pH = 8.5-10.0 and 30°C. The addition of as little as 5-10 mg formaldehyde per liter to these solutions reduced the treatment time by as much as 40% to less than 2 hours and reduced the final level of cyanide as measured to less than 0.1 mg per liter. The thiocyanate measured at the end of the reaction was generally within 2-5% of the value of the initial sample (25-150 mg/liter). For samples containing 5-10 mg cyanide per liter as sodium and copper cyanide, 75-125 mg hydrogen peroxide per liter was adequate for reduction of cyanide to levels below 0.1 mg/l in 2 hours.

Cyanide gold extraction process effluents containing 8-25 mg acid-dissociable cyanide, 3-10 mg soluble copper per liter and 75-250 mg thiocyanate per liter, were treated with formaldehyde and hydrogen peroxide under conditions used for the synthetic process solutions. It was shown that the acid-dissociable cyanide could be reduced to levels below 0.1 mg/liter in these solutions in less than two hours.

No destruction of thiocyanate was found by the colorimetric iron method. Based on this work it is concluded that the addition of 5 mg formaldehyde per liter and 75-100 mg hydrogen peroxide per liter to an effluent containing 4-6 mg soluble copper per liter and 5-20 mg acid-dissociable cyanide per liter is a good choice. The pH should be maintained between 9 and 10, and the temperature should preferably be between 20°C and 40°C to reduce cyanide levels to below 0.1 mg/l within 2 hours. The results obtained on mill effluents, some of which contained additional cyanide and copper (I) to cover anticipated operating ranges, are summarized in figures 7 and 8. The data are consistent with these obtained with synthetic solutions.

Figure 7

Lab Tests – Mill Water – I

Feed: 18 mg/l CN^-
147 mg/l CNS^-
36 mg/l Cu

mg/l

Feed		Effluent		
H_2O_2	HCHO	CN^-	CNS^-	Time (min)
146	-	0.04	145	111
145	5.0	0.11	144	79
96	-	0.06	143	176
87	5.0	0.07	141	119

Figure 8

Lab Tests – Mill Water – II

Feed: 147 mg/l CNS^-
3.6 mg/l Cu
5.0 mg/l HCHO

mg/l

Feed		Effluent		
CN^-	H_2O_2	CN^-	CNS^-	Time (min)
18	87	0.07	141	119
8	89	0.06	146	19
8	42	0.08	142	40

Addition of only 5 mg formaldehyde per liter to a natural mill effluent provided rapid destruction of acid dissociable cyanide to levels below 0.1 mg per liter in less than two hours. Figure 7 shows that cyanide destruction to levels below 0.1 mg per liter can be achieved even when the copper present in a mill effluent is much higher than required for efficient process operation.

Kinetics

It is important to provide sufficient time for the reaction to proceed as shown in figures 9 and 10. The rate of destruction of cyanide is increased towards the end of this reaction because of the increase in the ratio of hydrogen peroxide to residual cyanide. This can be explained by the kinetic rate equation derived by Masson[1]:

$$\frac{d\left[(CNO^-) + (CO_3^{-2})\right]}{dt} = k\,(H_2O_2)\,\left[(CN^-) + (\text{formate})\right]$$

An excess of 40-70 mg hydrogen peroxide per liter over that required for the oxidation of solution components has been sufficient to provide an effluent containing less than 0.1 mg acid-dissociable cyanide per liter.

Figure 9

Cyanide Gold Extraction Process Effluent

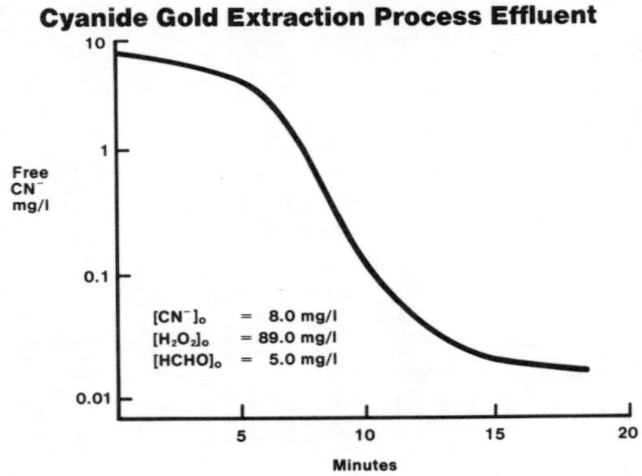

Figure 10

Cyanide Gold Extraction Process Effluent

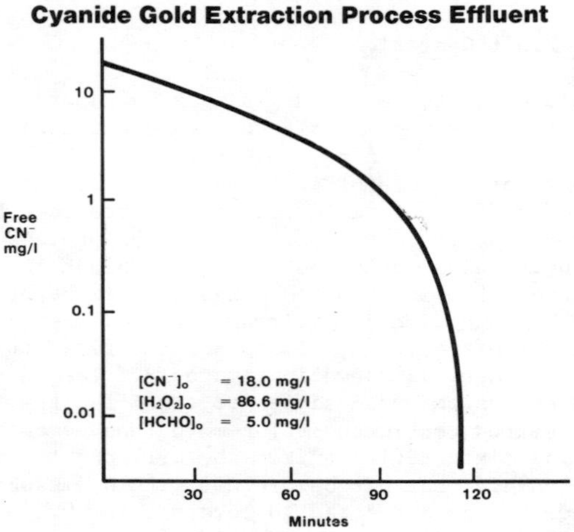

Treatment of both synthetic and real effluents with Kastone® peroxygen compound with or without formaldehyde yielded a precipitate that contained a mixture of basic salts that contained copper, iron, calcium, nickel and zinc salts that were quite easy to filter. Samples were analyzed for copper, iron and nickel by atomic absorption spectroscopy. Copper and nickel levels obtained were less than 0.1 mg/liter and iron levels were reduced to less than 0.3 mg/liter. The data in figure 11 shows typical data obtained in lab tests with synthetic solutions.

Figure 11

Filtered Effluent

	mg/l	
	CN^-	Cu
Lab	0.04	0.12
Pilot Plant	0.15	0.15

Figure 12

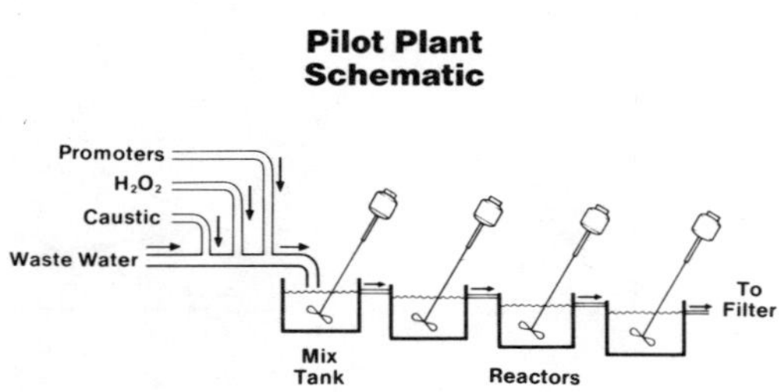

Pilot Plant Operation

This work has been done using a batch treatment concept. Based on experience with cyanide waste treatment installations, this chemistry can be scaled up on a batch basis or in a continuous treatment process if sufficient residence time is provided to allow the reaction to go to completion. The process is currently being evaluated in a pilot plant located at the Homestake Mining site in Lead, South Dakota. It has a 9-stage, cascaded, stirred-reactor design. A schematic of the feed, mixing and the first three reactor stages is shown in figure 12. The cascaded reaction system was designed to treat 1-2 gpm. It has operated continuously for more than 4 months treating mixed effluent from the mill and the mine. Filtered effluent has been clear. Acid-dissociable cyanide has been reduced to levels below 0.2 mg per liter but has not yet reached 0.05 mg per liter. Since mill effluent has been treated in the laboratory to give lower levels of acid-dissociable cyanide than that obtained from the pilot plant, it is concluded that a component in the mine effluent is a contributing factor to the higher cyanide content (figure 11). Work on this problem is in progress.

Analytical

During this study a simplified process control technique was developed. A silver/sulfide ion-specific electrode was used to monitor the change in cyanide ion concentration in the solution. A change in concentration by a factor of 10 changes the electrode potential (vs. a reference electrode) by about 118mV. This electrode has been particularly useful in determining when the reaction between the cyanide and the hydrogen peroxide is rapid and when it has gone to its final level. Reflux distillation analysis generally gave higher values of measured cyanide than the free cyanide ion calculated from direct electrode measurements. We conclude that the silver/sulfide ion-specific electrode is a useful process monitoring tool, but actual cyanide concentration in treated water should be measured in samples taken from the process and analyzed by tested methodology.

Conclusion

Copper-formaldehyde promoted hydrogen peroxide selectively destroys cyanide in the presence of thiocyanate in a "clear" gold milling process effluent. Laboratory tests have given clear filtered effluents containing less than 0.05 mg acid-dissociable cyanide per liter and 0.12 mg dissolved copper per liter. Pilot plant operation with mixed gold mill and mine effluent has given filtered effluents containing 0.15 mg acid-dissociable cyanide per liter and 0.15 mg dissolved copper per liter. The clarity of the pilot-plant filtered effluent has been below 5 NTU turbidity units.

Acknowledgements

We would like to acknowledge the participation of C. Sharpe Beltsch, R. Ommen and L. Trautman of the Homestake Mining Company at Lead, SD, for their contributions to the laboratory programs and for construction and operation of the pilot plant.

Bibliography

(1) O. Masson, J. Chem. Soc. 91, 1449-74 (1907).

(2) J. Zumbrunn, AITE International Conference of Nov. 15, 1968.

(3) B. C. Lawes, L. B. Fournier and O. B. Mathre, "Plating", Sept., 1973.

(4) U. S. Patent 3,617,567.

(5) U. S. Patent 3,617,582.

Chapter 10

THIOUREA AS A LIXIVIANT FOR GOLD AND SILVER

J. Brent Hiskey
Process Technology
Kennecott Minerals Company
Salt Lake City, Utah 84147

ABSTRACT

Thiourea is a reagent capable of complexing gold and silver. With the proper oxidant, thiourea can effectively dissolve metallic gold and silver. A thiourea leaching system possesses potential advantages over conventional cyanidation including acidic solutions instead of alkaline, cationic complexes in contrast to metal-cyanide anoionic complexes, and faster leaching rates with acidic solutions of thiourea. Generally, the rate of metal dissolution in thiourea solution is fastest when ferric ion is employed as an oxidant. Acidic thiourea solutions for certain applications could offer a nontoxic and noncorrosive option for processing gold and silver.

This paper provides information regarding the fundamental chemistry of the acidic thiourea system, and reviews the dissolution of both silver and gold in acidic solutions of thiourea. The recovery of gold and silver from acidic thiourea solution is discussed.

Introduction

The recent surge of gold and silver prices to record high levels justifies a fresh look at processing alternatives. Hydrometallurgical processes are especially suitable for the treatment of gold and silver bearing materials. To date the hydrometallurgical recovery of gold and silver has been established mainly upon cyanide leaching because of its economy and process simplicity. Other processes of minor significance have utilized sulfite and halide leaching chemistry. Thiourea represents another reagent capable of complexing gold and silver. Although thiourea leaching of gold and silver was discussed as long ago as 1941, only recently has it attracted serious support as a possible reagent for extracting gold and silver.

Potential applications of thiourea in the processing of gold were reviewed by Groenewald (1977). In that paper some basic aspects regarding the chemistry of the gold-thiourea system were discussed. In addition to the oxidative dissolution of gold in acidic thiourea solution, Groenewald critiqued other hydrometallurgical applications of thiourea in gold processing including solvent extraction, ion exchange, electroetching and electropolishing, electroless plating, and electrowinning.

General Aspects of Thiourea

Thiourea (H_2NCSNH_2) is an organic compound, the crystals of which dissolve in water to yield an aqueous form stable in acidic solutions. The unique feature of thiourea is that, in aqueous solution it reacts with certain transition metal ions to form stable cationic complexes. For example, the following generalized reaction illustrates the complexing role of thiourea.

$$Me^{n+} + x\ (Thio) \overset{\beta_x}{\rightleftarrows} [Me(Thio)_x]^{n+} \quad (1)$$

where Me is the metal; n, valence; Thio, represents thiourea; x coordination or stoichiometry number; and β_x, overall formation constant.

To understand the dissolution chemistry of gold and silver, it is helpful to establish the various equilibria associated with the thiourea system. Unfortunately, there is very little thermodynamic data for aqueous thiourea. The $\Delta G^\circ_{f,298}$ for thiourea was calculated from the limited information available from various sources. The solubility of thiourea as a function of temperature is shown in Figure 1. At 298°K, a saturated solution will contain approximately 142 g/l $CS(NH_2)_2$. The heat of solution at 298°K for $CS(NH_2)_2$ in water is 5.4 kcal/mole. From this information it was possible to calculate the various thermodynamic properties of aqueous thiourea. Table 1 summarizes the thermodynamic data of thiourea for both the crystal and aqueous form.

The reactions of thiourea with Au^+ and Ag^+ are of particular interest; however, reactions with other metal ions assume importance because they have the potential of consuming excessive amounts of thiourea. The nonferrous metals (Pb, Zn, Cd, Cu, etc.) are important in this regard. Furthermore, Fe^{3+} and Fe^{2+} thiourea complexes are of consequence in some leaching systems. In a mixed sulfate-thiourea system, the existence of both $[FeSO_4(CS(NH_2)_2)]^+$ and $[Fe(CS(HN_2)_2)]SO_4$ has been established.

Table 2 gives the overall formation constants of a number of metal thiourea complexes. The bis(thiourea)gold(I) ion is the only known thiourea complex of the aurous ion. The predominate thiourea complex of the argentous ion is the tris(thiourea)silver(I) ion, there is some evidence of the existence of the bis(thiourea)silver(I) complex.

There is abundant evidence that in the presence of some oxidants used to dissolve gold and silver, thiourea will undergo oxidation in various stages. The first oxidation product, formamidine disulfide, forms relatively easily in acid solution. Groenewald (1975) has studied the oxidation of thiourea on gold electrodes in acidic solutions. At anodic potentials greater than approximately 0.30 V, thiourea was oxidized to formamidine disulfide. The following redox reaction illustrates the first stage of thiourea oxidation.

$$2CS(NH_2)_2 \rightleftarrows [(NH_2)_2C\text{-}S\text{-}S\text{-}C(NH_2)_2]^{2+} + 2e \quad (2)$$

Succeeding reactions proceed at much slower rates, forming oxidation products that include elemental sulfur and even sulfate ions. Some oxidants (e.g., hydrogen peroxide) will oxidize thiourea quite fast at room temperature. Even though the oxidation potential for the Fe^{3+}/Fe^{2+} couple is sufficiently high to oxidize thiourea, the reaction between ferric ions and thiourea proceeds at a slow rate in sulfate solutions because of the stabilizing effect of the $[Fe(SO_4)(Thio)]^+$ complex.

There is considerable proof that formamidine disulfide behaves as an internal oxidizing agent. Tests where formamidine disulfide was formed in situ by the oxidation of thiourea with either dissolved oxygen or hydrogen peroxide revealed metal dissolution rates identical to those where formamidine disulfide was introduced from some external source.

In the absence of surface films, the overall rate of dissolution of gold and silver in thiourea solutions is in general controlled by diffusion of either reactants to the surface or products from the surface. Important in this regard are the diffusion coefficients for $Ag(Thio)_3^+$ and $Au(Thio)_2^+$. Shivrin and Shivrina (1976) have reported the following diffusivities at 298°K

$$D_{Ag(Thio)_3^+} = 4.7 \times 10^{-6}\ cm^2/sec$$

and

$$D_{Au(Thio)_2^+} = 3.3 \times 10^{-6}\ cm^2/sec$$

The diffusion coefficient of thiourea has been given as 1.1×10^{-5} cm^2/sec at 298°K (Gubailovskii, et al. 1972).

Application of Thiourea to Silver Leaching

Unlike base metals which are readily soluble in mineral acids, noble metals like gold and silver require the presence of oxidizing and complexing agents for dissolution. Silver approaches base metal behavior in acidic thiourea solutions ($Ag + 3Thio \rightarrow Ag(Thio)_3^+ + e$, $E° = 0.025$ V) requiring only mild oxidizing conditions for dissolution. Accordingly, the following reactions characterize the dissolution of silver in acidic thiourea with various oxidants

$$2Ag + 2H^+ + \tfrac{1}{2}O_2 + 6Thio \rightarrow 2Ag(Thio)_3^+ + H_2O \quad (3)$$

$$2Ag + H_2O_2 + 2H^+ + 6Thio \rightarrow 2Ag(Thio)_3^+ + 2H_2O \quad (4)$$

$$Ag + Fe^{3+} + 3Thio \rightarrow Ag(Thio)_3^+ + Fe^{2+} \quad (5)$$

Ferric sulfate and chloride have been frequently employed as oxidants. Other oxidants that appear compatible with acidic thiourea systems include Na_2O_2, O_3, $KBrO_3$, and $K_2Cr_2O_7$.

Like the metallic form, certain silver minerals require oxidizing conditions for dissolution in acidic solutions. Argentite (Ag_2S) in the presence of thiourea is leached by ferric ions according to the following overall reaction

$$Ag_2S + 2Fe^{3+} + 6Thio \rightarrow 2Ag(Thio)_3^+ + 2Fe^{2+} + S° \quad (6)$$

It is reasonable to believe that refractory silver minerals like Hessite (Ag_2Te) would respond to acidic thiourea leaching similar to the chemistry represented in Eq. (6).

Cerargyrite (AgCl) is a silver mineral that will dissolve in acidified thiourea solutions under nonoxidizing conditions. This mineral proceeds to solution by simple complexation dissolution. The following reaction represents the dissolution of AgCl in thiourea

$$AgCl + 3Thio \rightarrow Ag(Thio)_3^+ + Cl^- \quad (7)$$

This reaction is favored thermodynamically, at 25°C the equilibrium constant equals 2.2×10^3. Gubailovskii, et al. (1972, 1975) studied the kinetics of AgCl dissolution in aqueous solutions of thiourea. Their results suggest that the reaction is controlled by a diffusion mechanism. In acid solutions, thiourea decomposes with parting of elemental sulfur and formation of urea. The dissolution rate is limited by the dense sulfur film that coats the unreacted silver chloride. In a neutral medium (pH 5-7), silver sulfide forms a nonprotective layer on the silver chloride surface. This layer of silver sulfide gradually dissolves in the thiourea.

Early workers such as Plaksin and Kozhukhova (1941) used thiourea to dissolve gold and silver in H_2SO_4 and H_2O_2. They observed similar leaching behavior for both metals. Using solutions containing 10 g/l H_2SO_4 and 0.6 g/l H_2O_2, they found the dissolution to increase with increasing thiourea concentration in the range of 3-20 g/l $[CS(NH_2)_2]$. At these conditions the solutions were clear for up to 30 hr, after which noticeable amounts of elemental sulfur were liberated from the oxidation of thiourea. The addition of $FeCl_3$ effectively stabilized the thiourea for about 72 hr. At H_2SO_4 concentrations between 5 and 20 g/l, the dissolution was insensitive to sulfuric acid concentration.

More recently Plaksin and Kozhukhova (1960) studied the factors affecting the dissolution of gold and silver in solutions of thiourea. Various oxidizing agents were investigated; however, the rate was found to increase in the presence of Fe^{3+}. The dissolution rate increased linearly with increasing thiourea concentration in the range of 1-90 g/l $[CS(NH_2)_2]$. Optimum leaching results were achieved with a sulfuric acid concentration between 1 and 5 g/l.

Lodershchikov and co-workers (1975) investigated the kinetics of the dissolution of silver in aqueous solutions of thiourea. Although most of their data support a diffusion mechanism for leaching, some of the results were inconsistent with this interpretation. The complicating role of surface films at some conditions was believed responsible for these inconsistencies. It was demonstrated that thiourea is stable in the presence of ferric sulfate. The rate of silver dissolution increased linearly with increasing thiourea concentration until a maximum was reached, at which the rate was insensitive to further additions of thiourea. The maximum rates observed for 0.84 and 1.40 g/l $[Fe^{3+}]$ were 7.6×10^{-9} and 10×10^{-9} g-atom Ag/cm^2 sec, respectively. Figure 2 shows the dependence of the rate of silver dissolution on ferric ion concentration at 25°C for solutions containing 9.9 and 14.8 g/l thiourea. Both thiourea levels display two distinct kinetic regions. In the first kinetic region (i.e., low $[Fe^{3+}]$), the rate is controlled by diffusion of ferric ions and increases with increasing $[Fe^{3+}]$. The second region exhibits a decrease in rate because of a reduction in thiourea concentration (excess Fe^{3+} consumes thiourea by formation of Fe^{3+} thiourea complexes). The experimental activation energy was shown to be sensitive to thiourea concentration. At 1.12 g/l $[Fe^{3+}]$, the activation energy was determined to be equal to 9.61, 7.29, and 5.56 k-cal/mole for thiourea concentrations of 4.95, 12.2, and 22.8 g/l, respectively.

Ozone has been proposed as an oxidant for various hydrometallurgical applications. The dissolution of metallic silver in acidic thiourea by ozone was investigated by Chtyan, et al. (1975). Nearly complete (98%) dissolution of

-100 mesh Ag was achieved in 90 min using 55.9 g/l thiourea, 68.5 g/l H_2SO_4, and 0.06 g/l O_3.

Application of Thiourea to Gold Leaching

The dissolution of gold in acidic thiourea solutions as mentioned above was reported in the early work of Plaksin and Kozhukhova (1941, 1960). Since then, numerous investigators have examined the use of thiourea in considerable depth. Tataru (1968) determined the optimum conditions for the dissolution of granular gold in weakly acidic thiourea solutions. Hydrogen peroxide was used as the primary oxidant in this work. Using solutions containing 20 g/l HCl and 0.20 g/l H_2O_2, the dissolution results indicated that the optimum concentration range for thiourea was between 90 and 100 g/l. It was observed that ferric ion at low concentrations increased gold dissolution. The influence of the presence of Fe^{3+}, Cu^{2+}, Zn^{2+}, and Pb^{2+} on the rate of gold extraction was studied. Figure 3 shows the effect of these ions; the rate of gold dissolution is plotted as a function of the log metal ion concentration. In all cases, the rate of dissolution decreases with increasing metal ion concentration. Lead produces the most deleterious effect. Low concentrations of ferric ion enhance the rate; however, as the $[Fe^{3+}]$ increases, the rate falls off sharply. The thiourea system was applied to the extraction of gold from a flotation tailing, where gold was dispersed as ultramicroscopic or colloidal particles. Approximately 49% of the gold was extracted after 12 hr of leaching with a solution containing 90 g/l thiourea, 25 g/l HCl, and 2 g/l H_2O_2.

Lodeishchikov and co-workers (1968) used the thiourea-sulfuric acid-ferric ion system to extract gold from various ores. The optimum leaching conditions corresponded to concentrations of: 4 g/l thiourea, 5 g/l H_2SO_4, and 3 g/l $Fe_2(SO_4)_3$. Gold recoveries after 8 hr of leaching ranged from 92 to 98% depending on ore type. A preliminary acid washing was required to achieve maximum levels of gold recovery. Additions of poly(caprylamide) accelerated the leaching process. Similar results (Lodeishchikov, et al. 1972) were observed during the thiourea extraction of gold from cupreous gold-containing ores and concentrates.

Panchenko and co-workers (1975, 1976) studied the dissolution of gold, silver, and their alloys from a theoretical approach. Their results confirm that the dissolution of gold in excess thiourea is a first order process controlled by the rate of diffusion of oxidant (Fe^{3+}) to the surface. The rate of dissolution of the metals was found to be highest in the presence of $Fe_2(SO_4)_3$. The rate of gold dissolution was adversely affected by the presence of silver; this was attributed to the formation of a reaction inhibiting Ag-S film on the gold surface. Their work also indicated that the leaching of Ag_2S and $AgSO_4$ was faster in $CS(NH_2)_2$ than in NaCN solutions.

Groenewald (1976) has offered a comprehensive study of the dissolution of gold in acidic thiourea solutions. Sulfuric acid was employed exclusively during this investigation. However, various oxidants were examined including: oxygen, hydrogen peroxide, ferric ion, and formamidine disulfide. It was reported that Fe^{3+} produced the fastest initial rate of gold dissolution. At high thiourea concentrations and using Fe^{3+} as an oxidant, the dissolution rate approached diffusion control. It was concluded that ferric ion represented an impractical oxidant because of excessive consumption of thiourea by the formation of the $[FeSO_4(Thio)]^+$ complex. A major portion of the study was devoted to the oxidizing role of formamidine disulfide ion. Formamidine disulfide solutions were prepared by the in situ oxidation of thiourea from acid solutions. Dissolved oxygen and hydrogen peroxide were found equally effective as oxidants for the conversion process. Excess thiourea was initially added to compensate for the oxidation of thiourea. Formamidine disulfide appears to act as a direct oxidant for the dissolution of gold, since gold dissolves at the same rate in thiourea solutions when H_2O_2 and O_2 are employed as oxidants. In addition, identical results were obtained for acidic solutions of thiourea when a salt of formamidine disulfide was introduced as the oxidant.

Untreated and acid-washed samples of a crush gold ore were leached with acidic thiourea solutions. In general, the untreated samples exhibited a lower rate of gold dissolution than the acid-washed ore. This observation was attributed to the presence of soluble impurities. The influence of thiourea concentration was studied using the acid-washed ore sample. Using 0.05 M H_2SO_4 and a 6-hr air agitation leach, the level of gold extraction was 20, 50, and 90% for thiourea concentrations of 0.01, 0.03, and 0.10 M, respectively. For the same sample, very fast rates of gold dissolution were obtained using higher reagent concentrations (1.0 M H_2SO_4, 0.1 M H_2O_2, and 1.2 M thiourea). At these conditions, about 94% of the gold was dissolved in 15 min. Unfortunately, these impressive results are overshadowed by an extremely high consumption of thiourea. The author concluded that the thiourea technique was uneconomical when compared to traditional cyanidation approaches.

A recent paper by Chen, et al. (1980), discusses the leaching of gold and silver by "acidothioureation." Figure 4 compares acidic thiourea and cyanide leaching of gold and silver. For the conditions employed and up to 60 min, the rate for the thiourea system is approximately an order of magnitude greater than cyanide for both gold and silver. A copper concentrate containing 1.5 oz/ton (51.4 g/tonne) gold, 7.3 oz/ton (250 g/tonne) silver, and 6.02% Cu was leached with both thiourea and cyanide solutions (same solutions as given in Figure 4, except 0.5% CaO). The recovery of both Au and Ag was slightly higher with thiourea after 6 hr of leaching; interestingly, the level of copper dissolution was significantly lower with the thiourea system.

Ozone has been employed as an oxidant in the acid thiourea leaching of gold (Chtyan, 1976; Chtyan and Babayun, 1977). At 25°C the rate of gold dissolution was highest for a solution containing 61 g/l $CS(NH_2)_2$, 0.06 g/l O_3, and 10 g/l H_2SO_4. It was possible to extract about 98% of the gold from a concentrate in 200 min using an ozone-air mixture as an oxidant. In general, successful leaching where aggressive oxidant like hydrogen peroxide and ozone are used requires relatively high concentrations of thiourea. These oxidants consume the thiourea by the rapid oxidation to products where sulfur has a higher oxidation state. In contrast, ferric ion requires lower concentrations of thiourea for optimum leaching.

Recovery of Gold and Silver from Thiourea Solutions

The stability and cationic nature of the metal-thiourea complex makes the selective recovery of gold and silver by multiple methods possible. Most investigators have examined and developed techniques for the recovery of gold from acidic thiourea solutions. The following discussion, in the main part, reflects information assembled for gold recovery, in many cases the general results apply equally to silver recovery.

Lodeishohikov and Panchenko (1968) studied the use of activated carbon for the sorption of gold from acidic thiourea solutions. They observed gold loadings reaching 15 to 17% by weight of activated carbon. In addition, this work showed that activated carbon is effective in the sorption of the silver(I)-thiourea complex.

The solvent extraction of gold(I)-thiourea from solution has been accomplished using several reagents. Chernyak, et al. (1977) determined the optimum conditions for the recovery of gold from thiourea-hydrochloric acid solutions with tributyl phosphate (TBP) as an extractant. Tributyl phosphate has been used in a flotation technique (Bobrova, 1972) for the selective extraction of gold from hydrochloric acid thiourea solutions. The extraction of gold was observed to increase from 67% at pH1 to 99% at pH 8. An increase in the process time increased the level of extraction as well as the distribution coefficient. A variation of this technique employed a polyurethane foam (Braun and Farag, 1973 and 1973) loaded with TBP to concentrate gold. The solvent extraction of gold with bis(2-ethylhexyl) phosphate (Balakirava et al., 1975) from acidic thiourea solutions has also been investigated.

Certain strongly acidic cation-exchange resins will extract gold from acid thiourea solutions, whereas weak acid resins are inadequate for recovering gold in the presence of thiourea. Exchange fibers (Druzhina, et al., 1975) synthesized from polyvinyl alcohol will efficiently adsorb gold from both H_2SO_4- and HCl- thiourea solutions the exchange capacity was 15 to 23 mg Au/g fiber, silver is adsorbed to a lesser degree. Chernyak and coworkers (1977) have studied novel inorganic ion exchangers for recovering gold from thiourea solutions. They examined compounds like zirconium phosphate and molybdenum ferrocyanide as potential exchangers. For example, the capacity of the Mo ferrocyanide was 85 mg Au/g exchanger. Concentrated acid thiourea solutions will elute the gold; however, the sorption capacity decreases after Au desorption.

Numerous studies (Tataru, 1974 and 1974; Kocheva and Gacheva, 1973; Bek and Mal'tsova, 1972; Zdorova, et al., 1971) have been carried out using thiourea solutions to elute Au from resins saturated with gold-cyanide complex. This porcedure makes possible the recovery of the precious metals and allows the alkaline cyanide solution to be recycled after absorption of HCN with NaOH solutions. The precious metals are subsequently recovered by cementation or electrolytic precipitation. A variation of this technique involves the electroelution of gold and silver from ion exchangers, whereby thiourea solutions are employed for the simultaneous elution and electrodeposition of precious metals.

The reduction of gold(I)-thiourea to metal by electrodeposition methods has been suggested (Maslii and Zadonskaya, 1970; Bek, et al., 1969; Tataru, 1968). The electrolytic process can be accomplished in the presence of thiourea; however, sulfur formed following the decomposition of thiourea (anodic oxidation of thiourea) will contaminate the cathode. Tataru (1968) found that good agitation of the electrolyte and operating at current densities less than 250 amp/m^2 will eliminate this problem to a degree. Using a diaphragm cell where anodic and cathodic compartments are separate will eliminate sulfur contamination of the cathode and will avoid spontaneous dissolution of the cathode by oxidants formed by reaction at the anode. The ability to produce a pure gold product from thiourea solutions has not been commercially demonstrated.

Conclusions

This paper has attempted to furnish information relating to the chemistry of thiourea. Although the use of thiourea as a lixiviant for gold and silver was discussed over 40 years ago, just recently renewed interest in this reagent has become apparent. The following general conclusions were found regarding thiourea processes for gold and silver.

1. The dissolution of both metallic silver and gold in acidic thiourea medium is much faster than in cyanide solutions.
2. Chemical oxidants compatible with acidic thiourea solutions include ferric ion, hydrogen peroxide, ozone, and oxygen.
3. The rate of metal dissolution, especially gold, is fastest in the presence of ferric ion. However, ferric ion consumes excessive amounts of thiourea in the formation of iron-thiourea complexes.
4. The leaching of refractory forms of gold and silver where an agressive oxidant (e.g., H_2O_2) is required is viewed impractical because of the rapid oxidation (decomposition) of thiourea.
5. The application of thiourea for the treatment of precious metal-containing ores and concentrates is, for many cases, considered uneconomic because of excessive reagent consumption. Thiourea is often consumed by the formation of complexes with impurity metals, and by oxidation to products which are ineffective in complexing gold and silver.

6. The selective concentration and purification of gold and silver from thiourea solutions has been demonstrated for a variety of systems, none of which have been developed into a commercial recovery scheme.
7. Ion exchange stripping processes involving gold(I)-cyanide and gold(I)-thiourea systems possess distinct possibilities for the processing of gold.
8. Electrowinning of gold and silver from thiourea solutions requires a diaphragm cell to separate anodic and cathodic processes. This eliminates sulfur contamination of the cathode and avoids attack of the cathode by oxidants formed at the anode.

References

1. Balakireva, et al., "Solvent Extraction of Gold Using Bis (2-ethyl hexyl) Phosphate," Zh. Anal. Khim., 30(4), (1975), pp. 684-688. (CA 83:137,577b).

2. Bek, R. Uy., Maslii, A. I., and Lavrova, T. A., "Electrode Processes During the Electorlytic Removal of Gold from Thiourea Solutions," Tsvėt. Metal., 42(9), (1969), pp. 69-73.

3. Bek, R. Y. and Mal'tseva, A. B., "Recovery of Gold and Silver Present Simultaneously from a Thiourea Eluate," Izc. Sib. Otd. Akad. Nauk. SSSR, Ser. Khim. Nauk, 6, (1972) pp. 123-124.

4. Bobrova, A. S. , et al., "Flotation Extraction of Gold from Hydrochloric Acid Thiourea Solutions," Nauch. Tr., Irkutsk. Nauch.--Issled. Inst. Redk. Tsvet. Metal., 27, (1972), pp. 119-121. (CA 79:45,310z).

5. Braun, T. and Farag, A. B., "Reversed-phase Foam Chromatography. Chemical Enrichment and Separation of Gold in the Tributyl Phosphate-Thiourea-Perchloric Acid System," Anal. Chim. Acta, 65(11), (1973), pp. 115-126. (CA 79:38,168e).

6. Braun, T. and Farag, A. B., "Recovery of Gold from Thiourea Solutions with Open Cell Polyurethane Foams," Anal. Chim. Acta, 66(3), (1973), pp. 419-426. (CA 79:152,512n).

7. Chen, C. K., et al., "A Study of the Leaching of Gold and Silver by Acidothioureation," Hydrometallurgy, Vol. 5, 1980, pp. 207-12.

8. Chernyak, A. S., Bobrova, A. S., and Skobeer, I. K., "Gold Extraction from Chloride Thiourea Solutions," Anal. Technol. Blogorod. Metal., (1977) pp. 358-363. (CA 77:104,237p).

9. Chernyak, A. S., et al., "Sorption of Gold by Inorganic Ion Exchangers from Thiourea Solutions," Zh. Prikl. Khim., 50, (1977), pp. 1783-1786. (CA 87:155,292s).

10. Chtyan, G. S., et al., "Possibility of the Extraction of Metallic Silver in an Acidic Thiourea Solution Influenced by Ozone," Arm. Kbim. Zh., Vol. 28, No. 1, 1975, pp. 16-19. CA 83:31353k.

11. Chtyan, G. S., and Babayan, G. G., "Recovery of Gold and Silver in an Acid Thiocarbamide Solution under the Action of Ozone," Mater.-Vses. Mezhuuz. Konf. Ozoner, 2nd, 1977, pp. 17-18. CA 89:150,099r.

12. Chtyan, G. S., "Possibility of Extracting Gold Using an Acid Solution of Thiourea in the Presence of Ozone. II," Arm. Khim. Zh., Vol. 29, No. 3, 1976, pp. 225-9. CA 85:167,326c.

13. Druzhina, G. Ha., et al., "Solvent Extraction of Gold from Thiourea Solutions with Ion-Exchange and Electron-Transfer Exchange Fibers," Tsvetn. Met., 8, (1975), pp. 86-87. (CA 84:33,952z).

14. Groenewald, T., "The Dissolution of Gold in Acidic Solutions of Thiourea," Hydrometallurgy, Vol. 1, 1976, pp. 277-90.

15. Groenewald, T., "Electrochemical Studies on Gold Electrodes in Acidic Solutions of Thiourea Containing Gold(I) Thiourea Complex Ions," J. Appl. Electrochem., Vol. 5, 1975, pp. 71-8.

16. Groenewald, T., "Potential Applications of Thiourea in the Processing of Gold," J. S. Afr. Inst. Min. Metall., Vol. 77, No. 11, 1977, pp. 217-23.

17. Gubailovskii, V. V., et al., "An Examination of the Kinetics of Silver-Chloride Dissolution in Aqueous Solutions of Thiourea," Tsvet. Metal., Vol. 45, No. 5, 1972, pp. 36-37.

18. Kakovskii, I. A., and Gubailovskii, V. V., "Kinetics of the Dissolution of Silver Chloride in Aqueous Solutions Containing Chloride Ions," Soviet J. Nonferr. Metals Res., Vol. 3, No. 1, 1975, pp. 34-6.

19. Kocheva, L. L., and Gacheva, R., "Desorption of Small Quantities of Chloroaurate Ions Retained on Anion-Exchangers, in Separation of Gold from Base Metals," Talanta, 20(9), (1973), pp. 910-913.

20. Lodeishchikov, V. V., et al., "Investigation of the Kinetics of the Dissolution of Silver in Aqueous Solutions of Thiouréa," Izv. Vyssh. Uchebn. Zaved., Tsvetn. Metall., Vol. 2, 1975, pp. 77-81.

21. Lodeishchikov, V. V., et al., "Use of Thiourea as a Solvent for Ore Gold," Nauch. Tr., Irkutsk Gos. Nauch. - Issled. Inst. Redk. Tsvet. Metal., No. 19, 1968, pp. 72-84. CA 71:15,231r.

22. Lodeishchikov, V. V., et al., "Use of Thiourea for Extraction of Gold from Cupreous Gold-Containing Ores and Concentrates," Nauch. Tr., Irkutsk. Nauch.-Issled. Inst. Redk. Tsvet. Metal., No. 27, 1972, pp. 100-8. CA 79:56,306.

23. Loderschchikov, U. V. and Panchenko, A. F., "Using Activated Carbon for the Sorption of Gold from Acid Solutions of Thiourea, Soviet J. Nonferr. Metals Res., 9 (4), (1968) pp. 33-35.

24. Maslii, A. I., and Zadonskaya, A. V., "Kinetics of the Electrolytic Separation of Gold from Thiourea Solutions," Izv. Sib. Otd. Akad. Nauk SSSR, Ser. Khim. Nauk, 5(12), (1970), pp. 61-65.

25. Panchenko, A. F., et al., "Theoretical Basis of the Dissolution of Gold, Silver, and Their Alloys in Acid Thiocarbamide Solutions," Tezisy Dokl.-Vses. Sqveshch. Khim., Anal. Tekhnol, Blagorodn. Met., No. 1, 1976, p. 14. CA 90:12,879.

26. Panchenko, A. F., et al., "Kinetics of Dissolution of Gold, Silver, and Their Alloys in Aqueous Solutions of Thiourea," Izv. Akad. Nauk SSSR, Met., Vol. 6, 1975, pp. 32-7.

27. Plaksin, I. N., and Kozhukhova, M. A., "The Solubility of Gold and Silver in Thiourea," Compt. Rend. Acad. Sci. USSR, Vol. 31, 1941, pp. 671-4, CA 35:561.

28. Plaksin, I. N., and Kozhukhova, M. A., "Dissolution of Gold and Silver in Solutions of Thiourea," Sb. Nauchn. Tr., Inst. Tsvetn. Metal., Vol. 33, 1960, pp. 107-19. CA 57:10,888e.

29. Shivrin, G. N., and Shivrina, E. M., "Kinetics of the Electrochemical Reduction of Thiourea Complexes of Metals," Soviet J. Nonferr. Metals Res., Vol. 4, No. 5, 1976, pp. 192-4.

30. Sillen, L. G., and Martell, . E., "Stability Constants of Metal-Ion Complexes," The Chemical Society, London, 1964.

31. Tataru, S., "Dissolution of Gold in Thiourea Solutions," Rev. Rown. Chim., Vol. 13, No. 7, 1968, pp. 891-7.

32. Tataru, S., "Ion-Exchange Process for Gold Recovery," Rev. Roum. Chim. 19(4), (1974), pp. 717-725.

33. Tataru, S., "Recovery of Gold from Cyanide Solutions by Ion Exchange," Proc. Int. Miner. Process. Congr., 10th, Ed. M. J. Jones, IMM, London, 1974.

34. Tataru, S., "Electrolysis of Gold in Thiourea Solutions," Rev. Roum. Chin., 13(2), (1968), pp. 185-189.

35. Zdorova, E. P., Fridman, I. D., and Ivanorskii, M. D., "Electro-elution of Gold and Silver from Anion Exchangers," Anal. Tekhnol. Blagorod. Metal., (1971), pp. 363-369. (CA 77:116,845h).

TABLE 1

THERMODYNAMIC DATA
OF THIOUREA AT 298°K

Formula	State	ΔH°_f (kcal)	S° (cal/deg)	ΔG°_f (kcal)
$CS(NH_2)_2$	c	-22.1	72.44	-8.77
$CS(NH_2)_2$	aq	-16.7	91.81	-9.14

TABLE 2

OVERALL FORMATION CONSTANTS OF
METAL THIOUREA COMPLEXES

Complex ion	Log β_x	References
$Au(Thio)_2^+$	21.96 (21.50)	Shivrin and Shivrina, 1976 (Groenewald,1976)*
$Ag(Thio)_3^+$	13.10	Sillen and Martell, 1964
$Cu(Thio)_4^{2+}$	15.4	Sillen and Martell, 1964
$Cd(Thio)_4^{2+}$	3.55	(Sillen and Martell, 1964)**
$Pb(Thio)_4^{2+}$	2.04	(Sillen and Martell, 1964)**
$Zn(Thio)_2^{2+}$	1.77	Shivrin and Shivrina, 1976
$FeSO_4(Thio)^+$	6.64	Groenewald, 1976

*Calculated from standard potential measured for $Au+2(CS(NH_2)_2 \rightleftarrows Au(CS(NH_2)_2)_2^+ + e$

**The values reported by Shivrin and Shivrina, 1976) for $Cd(Thio)_4^{2+}$ and $Pb(Thio)_3^{2+}$ are incorrect.

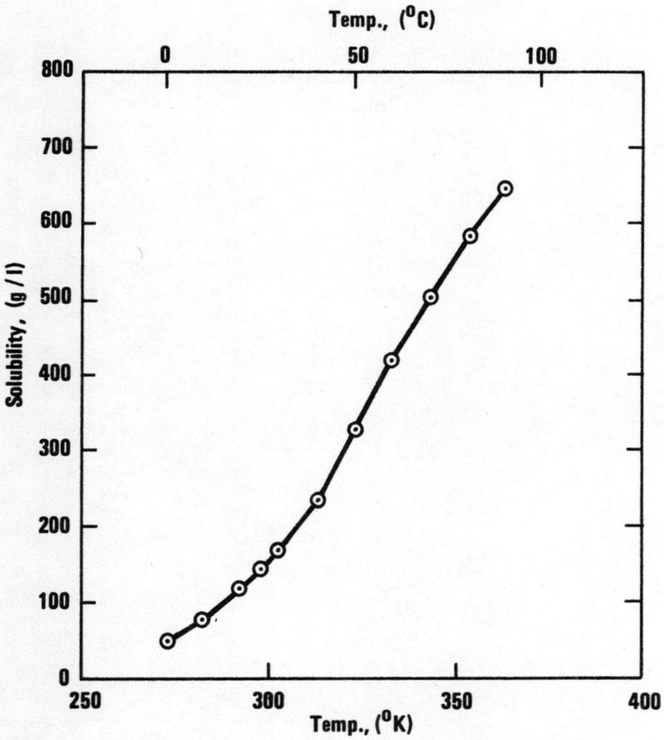

Figure 1. The Solubility of Thiourea in Water as a Function of Temperature

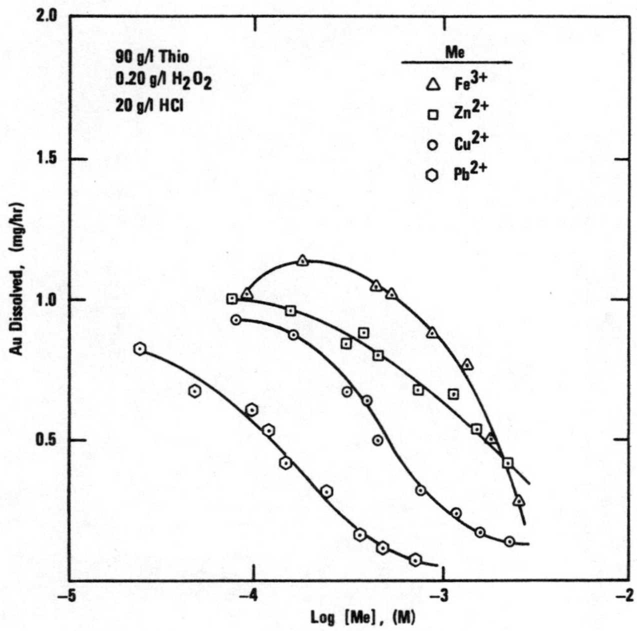

Figure 3. The Influence of Various Metal Ions on the Rate of Gold Dissolution in Acidic Thiourea Solutions

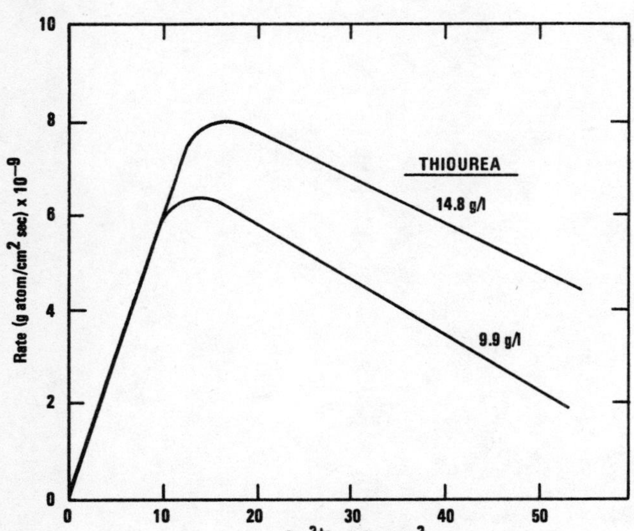

Figure 2. Dependence of the Dissolution Rate of Silver on Ferric Ion Concentration

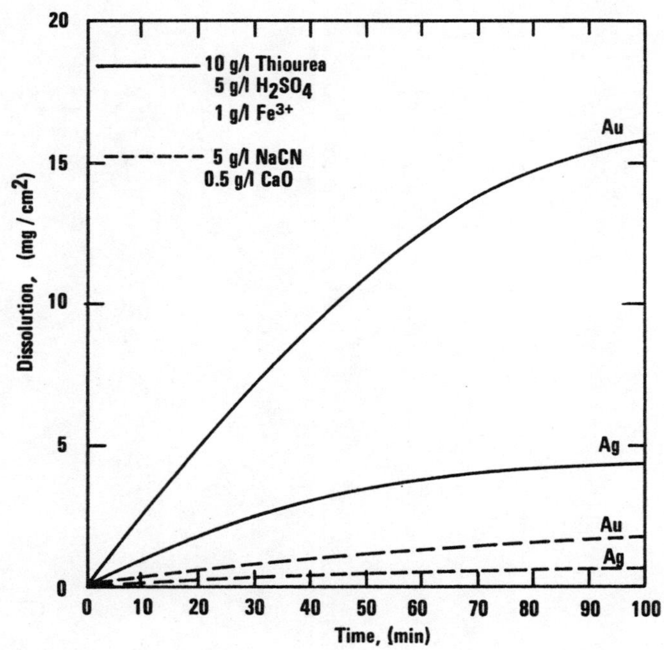

Figure 4. Dissolution of Gold and Silver from Rotating Disc (125 rpm) Using Acidic Thiourea and Cyanide Solutions

Chapter 11

EXTRACTION OF GOLD FROM A CARLIN–TYPE ORE USING THIOUREA

R. A. Pyper
Carlin Gold Mining Co.
Carlin, Nevada 89822

J. L. Hendrix
University of Nevada
Reno, Nevada 89557

ABSTRACT

Laboratory leaching experiments were conducted to ascertain the effectiveness of acidic solutions of thiourea, $CS(NH_2)_2$, for dissolving gold from finely disseminated gold ores. The ore tested was from the Chipmunk ore body of the Northumberland mining district in Nye County, Nevada, USA.

Extractions of up to 90% were attained from the ore. Extraction was shown to be dependent on thiourea concentration, ferric sulfate oxidant concentration, acid concentration, and leaching time. Optimum conditions were an ambient temperature leach near pH 1.0 utilizing 100 to 200 kg/ton thiourea and 50 kg/ton ferric sulfate for 8 to 24 hours.

INTRODUCTION

The use of thiourea, $(CS(NH_2)_2$ as a complexing/extracting agent for gold and other precious metals[1] has shown promise for implementation in the mining/metals industry. Laboratory testing of this reagent has indicated that a thiourea process for gold extraction has several advantages over conventional processing techniques. The use of cyanide for the extraction of gold from ores and concentrates is the predominant treatment method for primary recovery of this metal throughout the world. Indeed, cyanidation has proven to be very effective and economical for processing many types of gold-bearing materials. However, certain characteristics and disadvantsages of the cyanidation process warrant the investigation and development of alternative gold extractant systems.

The major concerns involving the cyanidation process are the health and environmental problems associated with it. These include the possible formation of deadly hydrogen cyanide gas, the ingestion or absorption through the skin of cyanide salts,[2] and the formation of the free forms of cyanide (HCN and CN^-) in effluent waters in concentrations toxic to aquatic life.[3,4] In light of increasing environmental awareness and responsibility, it is possible that continued stiffening of environmental and health standards will occur. Certainly, increased monitoring of air and water quality can be expected, while some states, such as California, may possibly ban new cyanide processing altogether.

Thiourea is not restricted by the same toxicity factors as cyanide. Thiourea toxicity data indicate high threshold limit values for mammals[5] and a lethal dose of 10 g/kg for humans.[6] On this basis, thiourea is to be considered much safer than cyanide. Indeed, the low pH required in a thiourea process would probably be of more concern. However, it must be mentioned that thiourea has been shown to be a carcinogen to rats and possibly to trout.[7] Even so, thiourea has been used for years in the treatment of thyroid diseases in humans and at this time is considered to be non-carcinogenic to humans.[7,8]

Another aspect of cyanidation which can be a problem is the somewhat slow kinetics of dissolution of gold by the cyanide species.[10,11] This can especially be a problem in vat and heap leaching operations which process ores containing fairly large-grained gold particles.

The kinetics of thiourea dissolution of gold have been studied extensively. Initial interest in thiourea as an extractant/complexing agent for gold dates back to a 1941 paper by Plaskin and Kozhukhova.[12] Interest was renewed by these researchers in 1960,[13] and since then several investigators have studied the gold-thiourea system.[1,9,10,14-24] Several of these investigators studied the rate of dissolution of pure gold in acidic thiourea solutions. Under proper conditions of acid and added oxidant, a thiourea medium has been shown to dissolve the gold as much as twelve times faster than a cyanide medium.[1,9,10,14,16,25] The acids studied included sulfuric, hydrochloric, and nitric. The various oxidants examined were ferric ion, hydrogen peroxide, sodium peroxide, formamidine disulfide, dissolved oxygen,[1] ozone,[17,18] and potassium permanganate.[19,20]. The fastest leaching system has been generally concluded to be the ferric ion-sulfate media system, although chloride media results are very similar. Rates are dependent on thiourea and oxidant concentrations, and appear to be controlled partially by chemical phenomena and partially by transport of oxidant species to the gold surface. A transition from diffusion control to kinetic control takes place at higher stirring speeds. These factors are similar to characteristics of cyanide dissolution of gold.[16]

With the exception of one study,[15] investigations on actual ground gold ores comparing cyanidation leaching rates to those of acidic thiourea solutions supported the results of the work on pure gold.[1,9,10] However, the increase in leaching rate was not as large as for pure gold samples. Other studies have been conducted on actual ores and/or concentrates[18,22,26] with recoveries of greater than 90% claimed using leaching times of eight hours or less.[18,22]

A third well-known disadvantage of cyanidation is the inhibition of the dissolution of gold caused by certain metal ions and other species.[11,27,28] These agents form from dissolution of constituents present in the ore. Such elements as copper and zinc tend to form ionic complexes with cyanide species, thus reducing the availability of cyanide ion for gold dissolution, as well as increasing cyanide consumption. These metal-cyanide species may also indirectly inhibit the gold dissolution reactions. Other elements, such as arsenic and antimony, react with cyanide solutions in such a way as to reduce the availability of dissolved oxygen necessary for the dissolution of gold.

Thiourea leaching has been reported to be affected to a much lesser degree by the presence of some of these cyanidation inhibitors.[9,10,13,16,22,23] In fact, one study suggests that copper, antimony, and arsenic minerals, and activated carbon actually increase the effectiveness of thiourea processing of gold.[22] However, other investigators studying thiourea in nitric and hydrochloric acid media report a decrease in solubility in the presence of zinc, copper, and lead salts due to complex formation of these metals with thiourea.[13,21] Iron minerals present may serve as a partial supply of ferric ion oxidant for the thiourea-gold interaction.[1,15] All in all, it has been concluded that thiourea has an advantage over cyanidation in the presence of these species.

The major disadvantage of the thiourea process is reported to be the higher cost due to high thiourea and acid consumptions.[1,9,15] Excessive thiourea consumption is partially due to the oxidation-degradation of thiourea by ferric iron. Other factors which contribute to thiourea degradation are temperature[15] and complex formation with other metals.[13,21] It has been shown that thiourea consumption can be reduced by an acid pre-leach to remove impurities.

Given the reported advantages of a thiourea process, it was decided to undertake groundwork experimentation on leaching of a Carlin-type ore. The parameters studied include thiourea concentration, oxidant requirements, pH, leaching time, and leaching temperature. Optimizations of overall gold extraction and thiourea usage were the goals of the experiments. The results are presented in this paper.

THEORETICAL CONSIDERATIONS

Thiourea in an acid medium forms a single cationic species with gold according to the following simple dissolution/complexation reaction:

$$Au^{o} + 2\ CS(NH_2)_2 = Au\ \ CS(NH_2)_2{}_{2}{}^{+} + e^{-} \qquad (1)$$

This reaction is considered reversible[14] with an as-written electrode potential (E^{o}) at 25°C of -0.38 v.[14,16] In the presence of an oxidizing agent such as hydrogen peroxide or ferric ion, thiourea will oxidize in successive stages to a number of products. The first step is the formation of formamidine disulfide:

$$2\ CS(NH_2)_2 = NH_2(NH)CSSC(NH)NH_2 + 2H^{+} + 2e^{-}. \qquad (2)$$

This is a rapid reaction with an electrode potential of +0.42 v. It has been shown that formamidine disulfide itself is a very active oxidant and is considered to be required for actual gold dissolution.[1] A combination of (1) and (2) yields the suggested overall equation:[16]

$$2\ Au^{o} + 2\ CS(NH_2)_2 + NH_2(NH)CSSC(NH)NH_2$$

$$+\ 2H^{+} = 2Au\ \ CS(NH_2)_2{}_{2}{}^{+} \qquad (3)$$

with an $E^{o} = +\ 0.04$ v and $G^{o} = -1845$ cal/mole (for two electron transfer).

In a sulfate medium containing ferric ion, oxidation is provided by the Fe^{+3}/Fe^{+2} couple:

$$Fe^{+3} + e^{-} = Fe^{+2} \qquad (4)$$

with $E^{o} = +0.77$ v. Combining (4) with (1) yields the following:

$$Au^{o} + Fe^{+3} + 2\ CS(NH_2)_2$$

$$= Au\ \ CS(NH_2)_2{}_{2}{}^{+} + Fe^{+2} \qquad (5)$$

with $E^{o} = +0.39$ v and $G^{o} = -8994$ cal.[17] A comparison of the free energies of (3) and (5) indicates a greater tendency for (5) to proceed; thus the presence of an added oxidant as well as formamidine disulfide is advantageous. It has also been suggested that a ferric ion-thiourea complex is formed which also aids in dissolution of gold.[9]

Continued oxidation of thiourea results in the slow formation of products in which sulfur has higher oxidation states. At alkaline pH values, oxidation products are urea and sulfate ion.[1] Formamidine disulfide has been shown to break down with time to cyanamide, hydrogen sulfide, and elemental sulfur.[1,15]

It has been noted that these secondary redox-reaction products tend to cause a passivation of the gold surface, thus partially reducing the rate of dissolution of gold.[1,9,12,14,16,21] However, the presence of ferric ion as an oxidant reduces the rate of these secondary reactions; thus, the most rapid rate of gold dissolution is obtained with ferric ion as oxidant.[1]

Thiourea tends to react slowly with ferric iron in a sulfate media forming very stable metal-ligand coordination bonds, such as:

$$Fe^{+3} + SO_4^{=} + CS(NH_2)_2 = FeSO_4(CS(NH_2)_2)^{+} \qquad (6)$$

This process continues with the reduction of the iron to the complex $Fe(CS(NH_2)_2\ SO_4$.[1] At increased thiourea concentrations, these reactions become more pronounced.[9]

With regard to pH, it was noted that the formamidine disulfide formation reaction (Equation (2)) is a reversible reaction, and the reduction potential value of +0.42 v is independent of pH from 0 to 4.3. However, it was also noted that the rate of dissolution of gold is strongly dependent on pH; i.e., increasing with decreasing pH.

METHODS AND MATERIALS

Ore Materials

The experimentation utilized material from the Chipmunk ore body of the Northumberland mining area in Nye County, south-central Nevada, USA. The ore is a finely-disseminated gold-bearing ore of the "Carlin" type as described by Wargo:[29]

"The classical Carlin-type gold deposit, as exemplified at the Carlin mine, contains micron-sized, high-fineness gold particles in a dolomitic siltstone. The mineralization is disseminated in the sense that it occurs more or less randomly in a block whose diameter might be a centimeter, a meter, or up to a kilometer. Randomly, however, does not mean accidentally. The gold occupies specific sites based on the chemistry of deposition and the nature of the structures present at the time of mineralization."

These types of deposit occur at various locations in Nevada and Utah[29,30,31] and "appear to be associated with hydrothermal processes related to late Tertiary tectonism and volcanism."[30] The Chipmunk ore is a quartz matrix with minor amounts of dolomite and trace muscovite. It contains 2.4 grams of gold per ton of ore (0.07 tr. oz Au/Avoir. ton rock) as determined by fire assay techniques. A semi-quantitative mass spectrographic analysis of the ore is presented in Table 1. The somewhat charcoal-colored appearance of the ore indicates the presence of some type of carbon material; however, this carbonaceous material is not activated and does not interfere with cyanide or thiourea extraction.

TABLE 1
SEMI-QUANTITATIVE SPECTROGRAPHIC ANALYSIS OF CHIPMUNK ORE

Element	Weight Percent	Element	Weight Percent
Ag	ND*	Na	ND
Al	3.0	Nb	ND
As	ND	Ni	0.004
Au	ND	P	ND
B	0.008	Pb	ND
Ba	ND	Pt	ND
Be	ND	Re	ND
Bi	ND	Sb	ND
Ca	2.0	Si	M**
Cd	ND	Sn	ND
Co	ND	Sr	ND
Cr	0.001	Ta	ND
Cu	0.001	Te	ND
Fe	7.0	Ti	0.1
Ga	ND	Tl	ND
Hf	ND	V	0.01
In	ND	W	ND
Mg	1.0	Zn	ND
Mn	0.06	Zr	0.007
Mo	ND		

* Not detected.
** Major Constituent.

TABLE 2
SCREEN ANALYSES OF CHIPMUNK ORE SAMPLES

Mesh	Microns	Weight Percent	Weight Percent Passing	Weight Percent	Weight Percent Passing
+48	297	27.0	73.0	7.3	92.6
65	210	10.2	62.8	10.6	82.0
100	149	7.4	55.4	11.9	70.1
200	74	12.6	42.8	14.0	56.1
-200	—	42.8	0	56.1	0
		100.0		99.9	

*Tyler screen mesh designation

The experimentation utilized two portions of the same sample. The screen analyses for the batches are presented in Table 2.

Analytical Methods

For solid samples, standard fire assay techniques[32] were employed to determine the gold content of head and tailing samples. Filtrate samples from thiourea extraction tests were analyzed by fire assay by evaporating 50 ml of solution in an assay-grade lead foil boat, combining with a fluxing mixture, and proceeding with a standard crucible assay. Liquid samples from cyanidation tests were also run in this fashion, but analytical checks were conducted using an atomic absorption spectrophotometer.

The concentration of thiourea in solution was determined by titration with 0.10 N mercuric nitrate solution using dephenylcarbazide as indicator.[33]

The concentration of free cyanide in solution was determined by titration with 0.025 N silver nitrate solution using para-dimethylamine-benzylidene rhodanine as indicator.

Testing Procedures

All of the cyanidation tests were conducted at room temperature using the standard bottle roll technique. The tests were designed to provide a standard gold recovery with which to compare the thiourea test results.

The majority of the thiourea test work was conducted at room temperature using the bottle roll technique to investigate the effects of pH, leaching time, thiourea concentration, and oxidant concentration on overall gold extraction and thiourea consumption. This method was basically the same as that for the cyanidation testing. For the thiourea tests, 100 grams of ore were placed in a 500-ml polyethylene bottle. A predetermined volume of 5 N sulfuric acid was then added to the bottle for conditioning the ore. At the end of the conditioning period, a known weight of reagent-grade thiourea was added along with a known weight of reagent-grade ferric sulfate (optional). Distilled water was added to obtain a slurry of approximately 30 wt.% solids. The ferric sulfate was supplied as an oxidant, but a large number of tests omitted the oxidant. The bottle test then commenced.

Agitated Open Beaker Tests

In order to investigate the effects of temperature on the efficiency of gold extraction with thiourea, several 24-hour tests were conducted in 600-ml open beakers. An impeller-type agitator with a variable speed electric motor was used to provide mixing. This arrangement also allowed for continuous monitoring of oxidation potential if desired. Temperature was regulated by either a water shaker bath or a cooling plate.

Acid Pre-Wash of Ore

A test was conducted in which 100 grams of ore was contacted with 500 ml of 3.57 N (16.3 wt.%) sulfuric acid for three hours. The ore was then filtered in the pressure filter, and the cake was weighed to obtain the amount of liquid hold-up. The cake was repulped to 200 ml of liquid with distilled water. Ten grams of thiourea were added and the bottle was run through the bottle roll test sequence with a 24-hour leach time.

Recycle of Leachant

In order to obtain information pertaining to re-use of leach liquor, as would be the case in a larger-scale continuous-type leaching operation, a test was run in which 50 ml of filtrate from a 24-hour bottle roll test were mixed in a bottle with 50 grams of fresh ore and 50 ml of 5 N sulfuric acid. The bottle was then agitated for 24 hours, followed by the standard bottle-roll treatment.

Two-Stage Leach

A two-stage leach was conducted as a further step toward simulation of continuous operation, as well as to provide insight into possible establishment of chemical equilibria in a one-stage test. This was accomplished by repulping the filter cake from a room-temperature open-beaker test (after 11 hours of leaching) with 10 ml of 5 N sulfuric acid, 140 ml of distilled water, and 10 grams of thiourea. The second slurry was leached at room temperature for 12 hours in an open beaker.

RESULTS

Cyanidation

Six 24-hour bottle-roll cyanidation tests were conducted to obtain a "standard" extraction to compare thiourea leaching test results with. Using a 10 kg/t sodium cyanide concentration yielded an average extraction of 83%. Sodium cyanide and calcium hydroxide consumptions were 1.0 and 0.6 kg/t, respectively. A one-hour test was also conducted at the same cyanide treatment level yielding an 81% extraction.

Thiourea

The effect of pH on the extraction of gold for three different thiourea concentrations was determined. The results are presented in Figure 1. The maximum extraction is 86% at pH 0.6 using the 35 kg/t thiourea concentration. At each concentration level the behavior is similar: increasing extraction with decreasing pH. Plateaus in the extraction are reached at approximately 62% near pH 2 for the 10 kg/t concentration, 82% near pH 3 for the 200 kg/t concentration, and 85% at pH 1 for the 35 kg/t concentration level. Sulfuric acid consumption was about 250 kg/t for the leach tests near pH 1.0.

The results of increasing thiourea concentration on gold extraction are presented in Figure 2. These 24-hour tests were run at three different sets of pH and oxidant leaching conditions. All three curves appear to level off above about 25 kg/t thiourea loading. The highest extraction (88%) was attained using 200 kg/t thiourea and 5 kg/t $Fe_2(SO_4)_3$. The other curves attain the 80 to 85% extraction level and remain there throughout the range of concentrations.

Figure 3 shows the change in the extraction of gold as related to the amount of $Fe_2(SO_4)_3$ oxidant used. The results at the different thiourea levels do not appear to form a definite pattern: recoveries at the 25 and 40 kg/t loadings tend to drop with increasing $Fe_2(SO_4)_3$ present, while the recovery at the 35 kg/t $Fe_2(SO_4)_3$ level and then falls off. The 100 kg/t level remains fairly constant at about 85%. The data at 200 kg/t thiourea also suggest an increase in recovery upon addition of 5 kg/t $Fe_2(SO_4)_3$. The addition of the $Fe_2(SO_4)_3$ increased the extraction from 81% to 88%.

The variation in gold extraction with respect to leaching time is observed in Figure 4. At the lowest thiourea concentration (10 kg/t data not plotted), the change in extraction is somewhat erratic as leaching time is increased. It is observed at 168 hours (7 days) of leaching, extraction appears to drop off from the 60 to 65% level of the shorter leach times to approximately 50%. At 35 kg/t thiourea and 5 kg/t $Fe_2(SO_4)_3$, the extraction increases steadily from 75% at 2 hours to 85% at 24 hours. Data at this thiourea level but without oxidant indicate similar trends with a leveling off of extraction at 86% up to and beyond 24 hours of leaching. At 100 kg/t thiourea, it is observed that 77% of the gold can be extracted in one hour and extraction continues to increase to 90-plus % at 48 hours. At the highest thiourea level of 200 kg/t, the extraction is constant at about 81% over the range studied (1 to 48 hours). No data are available at leaching times of less than one hour.

Much data were obtained on thiourea consumption as each of the previously mentioned parameters were varied. However, some difficulty arose in obtaining a highly accurate consumption number for any particular test. The main problem encountered was the obscuring of the titration end point due to the presence of large quantities of iron species, sulfates, and other salts. Also, the nature of the calculation results in large increases in consumption values for very small decreases in volume of titrant used. Coupling these two facts results in the necessity to downplay the specific consumption number. Instead, one must seek out any trends which may be established by comparison of consumption values which vary considerably for a small change in a given parameter.

The consumption data for several series of tests are presented in Table 3. Initially, it is noted that a strong pattern relating thiourea consumption to gold extraction is not clearly established. Thiourea consumption is plotted as a function of leaching time for three sets of initial thiourea concentrations in Figure 5. In all cases the consumption jumps considerably from the short leaching period values (0.5 to 2.5 kg/t) to the 24-hour values (9 to 11 kg/t). However, beyond 24 hours the results become erratic; i.e., at the 35 and 100 kg/t loadings the consumption appears to drop somewhat, while at the 200 kg/t level the consumption continues to increase to 17.5 kg/t as leach time continues to 48 hours.

The variation of thiourea consumption with amount of $Fe_2(SO_4)_3$ added is shown in Figure 6. It is observed that for these sets of conditions thiourea consumption increases with increasing oxidant concentration. The curve at the 40 kg/t treatment level appears to be leveling off above the $Fe_2(SO_4)_3$ loading at 50 kg/t, but there is insufficient data at the 200 kg/t thiourea level to check this effect.

The effect of acid level on thiourea consumption is shown in Figure 7 for two sets of leaching conditions. Since the pH effect on gold recovery had been established, further testing at higher pH values was not conducted. However, it does appear that the acid levels near pH 1.0 (100 to 120 ml/t H_2SO_4) have little effect on thiourea consumption.

Figure 8 presents plots of thiourea consumption and gold extraction as functions of leaching temperature for a single set of leaching conditions. Thiourea consumption is fairly constant up to 40°C, after which it rises considerably. The temperature effects on gold recovery are somewhat dramatic in that leaching above room temperature results in a considerable drop in recovery (90% at 21°C to 54% at 60°C).

TABLE 3
EFFECTS OF VARIOUS PARAMETERS ON THIOUREA CONSUMPTION: CHIPMUNK ORE

Series	Time (hr)	$Fe_2(SO_4)_3$ (kg/t)	Acid (ml 5N H_2SO_4)	Temperature (°C)	Thiourea (kg/t)	Consumption (kg/t)	% Extraction
Time	8	0	100	21	35	1.5	77
	24	0	100	21	35	9	96
	48	0	100	21	35	7	86
	1	0	100	21	100	0.5	73
	24	0	100	21	100	9.5	89
	48	0	100	21	100	6	92
	4	0	100	21	200	2.5	77
	24	0	100	21	200	11	80
	48	0	100	21	200	17.5	82
$Fe_2(SO_4)_3$	24	5	100	21	40	6.5	82
	24	50	100	21	40	14.5	71
	24	100	100	21	40	17	75
	24	0	100	21	200	2.5	75
	24	5	100	21	200	10	77
Acid	24	0	100	21	35	9	86
	24	0	120	21	35	10.5	82
	1	0	100	21	100	0.5	73
	1	0	120	21	100	0.5	81
Temperature	24	0	100	10	100	12.5	68
	24	0	100	21	100	9.5	90
	24	0	100	40	100	12.5	81
	24	0	100	60	100	22	54
Thiourea	24	0	100	21	35	9	86
	24	0	100	21	100	9.5	89
	24	0	100	21	200	11	80

TABLE 4
FINAL OXIDATION POTENTIAL VALUES FOR VARIOUS THIOUREA LEACHING CONDITIONS: CHIPMUNK ORE

Thiourea (kg/t)	$Fe_2(SO_4)_3$ (kg/t)	Acid (ml 5N)	Time (hr)	Temperature (°C)	mv	% Extraction	Comments
35	50	100	24	21	224	80	
35	0	100	48	21	191	86	
100	0	100	48	21	158	92	
100	0	100	24	21	160	90	
150	0	100	24	21	158	86	
100	0	100	24	21	174	83	Acid pre-wash
100	0	100/50	20/20	21	164/193	84/78	Two-stage
100	0	100	24	21	168	84	Continuous Monitor
100	0	100	24	10	144	68	
100	0	100	24	21	160	90	
100	0	100	24	40	NA*	81	
100	0	100	24	60	109	54	

*Not available.

The change in thiourea consumption with changing initial thiourea treatment level is presented in Figure 9 for one set of leaching parameters. It is observed that the consumption value is fairly constant (9 to 11 kg/t) over the thiourea concentration range studied.

Oxidation potential measurements were taken on the bottle roll test filtrates of the final portion of the test work, after some of the most effective leaching conditions were established. Also, oxidation potential was monitored continuously on several open beaker tests. Table 4 presents the final readings along with leaching conditions, and Figure 10 presents the oxidation potential change with leaching time at constant leaching conditions. Also plotted is gold extraction vs leach time data for the same experiments. The potential rises rapidly from +124 mv at 0.16 hours to +157 mv at 5 hours where the curve flattens between +158 and +170 mv; the extraction curve seems to follow the same pattern, although there is an acute lack of data between 1 and 20 hours of leaching.

It is noted that insufficient data have been collected to establish clear-cut patterns regarding the effects of the various parameters on oxidation potential and gold extraction. It does appear, however, that a potential above about +150 mv is required to obtain an extraction of 80 to 85%; but potentials higher than +150 mv do not result in a pronounced increase in extraction above 85%. Temperature data are also somewhat incomplete; however, it is noted that lower oxidation potentials for the 10 and 60°C tests coincide with lower recoveries as compared to room temperature tests.

In Table 5, the results of the acid pre-wash test are compared to identical tests without pre-wash. It is observed that the extraction and oxidation potential are fairly constant at about 85% and +170 mv, respectively. However, thiourea consumption for the pre-wash appears to be somewhat higher at 9 kg/t vs 6 kg/t, although this could be within the error margin of the analytical procedure.

TABLE 5
ACID PRE-WASH COMPARISON: CHIPMUNK ORE

Pre-wash conditions: Acid Solution - 3.57 N; Acid Volume - 500 ml; Wash Time - 3 hours; Weight of Ore - 100 g

Leach Conditions: Temperature - 21°C; Time - 24 hours; Thiourea - 100 kg/t; $Fe_2(SO_4)_3$ - 0; pH - 0.8

Test Type	Number of Tests	mv	Consumption (kg/t)	% Extraction
Standard	5	+164	6	87
Pre-Wash	1	+174	9	83

Results of the leachant recycle are presented in Table 6 in comparison with a 24-hour bottle roll test using similar leaching conditions. The first stage of the leach produced results nearly identical to those of the standard test. However, the second-stage recycle extraction is somewhat lower at 78% (vs 85%) while the oxidation potential jumps from +164 mv. Thiourea consumption is reduced to 2 kg/t vs 5 to 6 kg/t for the standard and first-stage tests.

TABLE 6
RESULTS OF LEACHANT RECYCLE: CHIPMUNK ORE

Leach Conditions: Temperature - 21°C; Thiourea - 100 kg/t/stage; $Fe_2(SO_4)_3$ - 0; pH - 0.9-1.0

Test Type	Leach Time (hr)	Acid Level (ml 5N H_2SO_4)	mv	Consumption (kg/t	% Extraction
Standard	24	100	+164	6	87
Recycle:					
Stage 1	20	100	+164	5	84
Stage 2	20.5	50	+193	2	78

The two-stage leach test results are compared to a standard test at similar conditions in Table 7. Total extraction by the two stages was similar to that of the standard test at approximately 85%. Thiourea consumption increased from 0.5 kg/t in the first stage to 6 kg/t in the second. Second-stage consumption was the same as that of the standard test.

TABLE 7
RESULTS OF LEACHANT RECYCLE: CHIPMUNK ORE

Leach Conditions: Temperature - 21°C; Thiourea - 100 kg/t/stage; $Fe_2(SO_4)_3$ - 0; pH - 0.8

Test Type	Leach Time (hr)	Acid Level (ml 5N H_2SO_4)	Consumption (kg/t	% Extraction
Standard	24	100	5	86
Recycle:				
Stage 1	11	100	0.5	81
Stage 2	12	10	6	3
Total	23		6.5	84

DISCUSSION

Cyanidation

The bottle roll cyanidation test results on Chipmunk ore indicate that cyanide leaching will yield an 80-plus % extraction even at short leaching times. This is probably due to the finely-disseminated nature of the gold particles in the ore. Although extensive test work was not conducted on the cyanidation aspect, it is noted that extractions approaching 90% were not attained.

Thiourea

The results shown in Figure 1 reveal that thiourea extraction is a function of pH; i.e., increasing extraction with decreasing pH until extraction levels off near pH 2. This is consistent with the gold complexation reaction (3), as well as results published by other investigators.[9,18,34] Although the highest extraction given in the table occurs at the 35 kg/t thiourea level, it is noted that higher thiourea concentrations allow a larger range of effective leaching pH's before extraction begins to drop off.

Gold extraction is also a function of thiourea concentration as evidenced by Figure 2. All three sets of data indicate that thiourea concentrations of greater than 50 kg/t are required to obtain an 80-plus % extraction. This effect is also consistent with the literature[1,9,12,13,16,18,34] as well as the gold complexing Reactions (1) and (3). The drop in recovery at the higher concentrations for the pH 1.0 series is possibly due to the breakdown of formamidine disulfide resulting in a sulfur coating of the gold particles as discussed in the section on theory.

Regarding the presence of ferric sulfate as oxidant (Figure 3), it is observed that in every case the highest extraction was obtained using 5 kg/t $Fe_2(SO_4)_3$. However, extractions of 80-plus % were obtained when no oxidant was added. It is, therefore, assumed that the majority of oxidant required is supplied by the ore itself. Indeed, the spectrographic analysis of Table 1 indicates that the ore is 7% iron. However, a small additional amount is required; but higher $Fe_2(SO_4)_3$ additions result in a slight drop in extraction. This could be due to the breakdown of formamidine disulfide in the presence of higher amounts of ferric ion, forming sulfur and other secondary redox reaction products which are known to passivate the gold surface.[1,9,12,14,16,21] Another explanation could be a depletion in the supply of thiourea resulting from complexing with ferric ion (Reaction (6)). The latter explanation could be especially applicable to the lower initial thiourea test work (25, 35, 40 kg/t).

The data on leaching time versus gold extraction, Figure 4, are interesting in that the reported kinetic factor is observed only at the higher thiourea concentrations. Indeed, comparison with the one-hour cyanidation extraction of 81% suggests that for this particular ore, thiourea may have no kinetic advantage at all. It is possible that at shorter leaching times (less than one hour) thiourea would extract gold noticeably faster than cyanide, but no data are available to demonstrate this. Lower extractions at the 10 kg/t thiourea level are probably due to a shortage of available thiourea. At the 35 and 100 kg/t levels the extraction seems to be a direct function of leaching time, while at the 200 kg/t level the rapid leveling of the extraction at 80 to 82% is again possibly due to passivation of the gold surfaces.

According to the literature, thiourea consumption should vary directly with ferric ion concentation, leaching time, temperature, and the presence of certain impurities.[1,9,13,15,21] Except for the latter factor which was not studied, the data in Table 3 and Figures 5 through 9 support these conclusions. The remaining three factors are closely related in that increased consumption is due to the ultimate breakdown of the formamidine disulfide to sulfur and other species which do not aid in gold complexation. Ferric ion will complex thiourea as well as aid in its degradation, thereby compounding its affect on consumption. Regarding leaching time, it would be expected that consumption would steadily increase as leaching proceeded. However, only at the 200 kg/t thiourea level is this effect noted past 48 hours. The absence of added oxidant in the 35 and 100 kg/t series could indicate that by 24 hours of leaching the ferric ion present in the ore has all been consumed (reduced to ferrous) and further thiourea degradation is slowed. At the 200 kg/t level, the presence of such a high excess of thiourea could contribute to a continuation of thiourea degradation. This may provide a clue to the relatively low extractions observed at the 200 kg/t level in the leaching time/extraction data.

The effect of leaching temperature on thiourea consumption and gold extraction as shown in Figure 8 indicates that thiourea is fairly stable in solution until temperatures rise above 40°C. The other consumption numbers are within experimental error. The drop in recovery at 60°C could possibly be due to thermal degradation of the gold-thiourea complex itself, as well as a loss of available thiourea/formamidine disulfide due to thermal effects.

Thiourea consumption does not vary with increasing initial thiourea concentration, maintaining a constant consumption of approximately 10 kg/t for 24-hour tests. Literature values vary between 1.4 and 5 kg/t, although leaching conditions are not fully specified.[9] It is observed that to obtain a reasonable gold rcovery (greater than 80%) from this ore requires either very high thiourea concentrations or leaching times of eight hours or greater. For the former case, thiourea consumption will be relatively high; a lack of thiourea consumption data for intermediate leaching times at lower thiourea concentrations prevents further discussion of these variables. It should be noted that high thiourea consumption coupled with the very high acid consumption of 250 kg/t could result in a rather expensive gold extraction process.

Study of oxidation potential data for leaching of Chipmunk ore indicates a potential of +150 to 160 mv is required for 80-plus % extraction of the gold. Potentials beyond this level, however, appear to be related to factors other than gold-thiourea-formamidine disulfide complexation.

The addition of an acid pre-wash to a 24-hour leach exhibited no advantage whatsoever (Table 5). Actually, a slight drop in recovery and an increase in thiourea consumption are observed. But it is suspected that these differences are within experimental error. Several literature sources mention the advantage of the use of a pre-wash to reduce thiourea consumption,[9,10,13,22] and it is curious that such an effect was not observed in this case. However, it is possible that a shorter leaching time would help to reduce thiourea consumption while still maintaining an acceptable extraction.

The leachant recycle experiment indicates that a thiourea solution can be used for successive leaches while maintaining respectable extraction. The drop in consumption from 5 to 2 kg/t is consistent with work done by others and is encouraging regarding continuous ore processing. This drop could possibly be due to the initial conversion of a portion of the thiourea to formamidine disulfide, which would register as a loss of thiourea but would still be effective in gold dissolution.

Results of the two-stage leach (Table 7) indicate that an increase in gold extraction above 85 to 90% may not be possible on this ore. It can be concluded that this extraction ceiling is not due to a solution equilibrium being established in the single stage batch tests. Rather, this ceiling effect could be due to gold surface passivation by redox reaction products. A further explanation could be that the fraction of gold that is not extracted is locked inside insoluble particles and thereby shielded from the thiourea species. This would explain the fact that cyanidation is also limited to about 80% extraction. The slight extraction advantage of thiourea would then be due to the dissolving effect of the acid treatment.

The increase in thiourea consumption in the second stage confirms the results of the acid pre-wash in that relatively high consumptions are still encountered even after pretreatment with an acid liquor. However, in this case the leaching time was shorter (12 hours vs 24 hours) and it is known that very little of the thiourea was tied up in gold complexation.

CONCLUSIONS

1. Thiourea leaching of Chipmunk ore is technically possible, with gold extractions equal to or exceeding extraction obtained by cyanidation. However, relatively high consumptions of thiourea and sulfuric acid may make thiourea processing of this ore expensive in comparison to cyanidation.

2. Due to the finely-disseminated nature of the gold in this ore, thiourea leaching does not exhibit a practical kinetic advantage over cyanidation.

3. Optimum leaching conditions for an 80-plus % gold extraction using thiourea ore:

Thiourea	-	50-100 kg/t
Ferric Sulfate	-	5 kg/t
Temperature	-	Ambient
Time	-	8 to 24 hours
pH	-	1.0
Oxidation potential	-	+160 mv

4. Prewashing Chipmunk ore with an acid solution does not yield advantages in thiourea consumption or gold extraction.

5. Re-use of thiourea leach liquors results in effective use of the thiourea, thus reducing thiourea consumption while maintaining adequate extraction levels.

ACKNOWLEDGEMENT

The authors wish to acknowledge the Cyprus Mining Company for allowing the results of this study to be presented.

REFERENCES

1. Groenewald, T., "Potential applications of thiourea in the processing of gold," J. S. Afr. Inst. Min. Metall., Vol. 77, No. 11, 1977, pp. 217-23.

2. National Institute for Occupational Safety and Health, Publication No. 77-108, 1976; "Occupational Exposure to Hydrogen Cyanide and Cyanide Salts," 191 pp.

3. Encyclopedia of Chemical Technology, Kirk, R. and Othmer, D. (eds.), 3rd Ed., Vol. 7, John Wiley and Sons, New York, 1979, p. 317.

4. Ecological Analysts, Inc., "Cyanide: An overview and analysis of the literature on chemistry, fate, toxity, and detection in surface waters." Report prepared for the Inter-Industry Cyanide Group, June 1979, p. 2-10.

5. National Institute for Occupational Safety and Health, Publication No. 79-100, 1978; Registry of Toxic Effects of Chemical Substances, p. 1285.

6. Sakai Chemical Industry Co., Ltd., "Material Safety Data Sheet for Thiourea." Supplied by McKesson Chemical Co.

7. International Agency for Research on Cancer (IARC), "Monographs on the Evaluation of the Carcinogenic Risk of Chemicals to Man: Some anti-thyroid and related substances, nitrofurans and industrial chemicals," Vol. 7, Feb. 1974, pp. 95-109.

8. Shubik, P., "Potential carcinogenicity of food additives and contaminants." Cancer Research, Vol. 35, Nov. 1975, pp. 3475-80.

9. Groenewald, T., "The dissolution of gold in acidic solutions of thiourea." Hydrometallurgy, Vol. 1, 1976, pp. 277-90.

10. Chen, C. K., Lung, T. N., and Wan, C. C., "A study of the leaching of gold and silver by acidothioureation," Hydrometallurgy, Vol. 5, 1980, pp. 207-12.

11. American Cyanamid Co., "Chemistry of Cyanidation," Mineral Dressing Notes, No. 23, Dec. 1978, p. 7.

12. Plaskin, I. and Kozhukhova, M., "The solubility of gold and silver in thiourea," Doklady Akadomii Nauk SSR, Vol. 31, 1941, pp. 671-4 (Russ).

13. Plaskin, I. and Kozhukhova, M., "Dissolution of gold and silver in solutions of thiourea," Sbornik Nauchuykh Trudov, Institut Tsvetnykh Metallov, Vol. 33, 1960, pp. 107-19 (Russ).

14. Groenewald, T., Electrochemical studies on gold electrodes in acidic solutions of thiourea containing gold (I) thiourea complex ions," Journal of Applied Electrochemistry, Vol. 5, 1975, pp. 71-8.

15. Mineev, G., Stroganov, G., and Syrtlanova, T., "Technological investigation of the filtration leaching of gold by acidic thiourea solutions." Translated from Zhurnal Prikladnoi Khimii, Vol. 50, No. 10, 1977, pp. 2189-92.

16. Panchenko, A., Kakovskiy, I., Shamis, L., and Khmelnitskaya, O., "Dissolution kinetics of gold, silver, and their alloys in aqueous thiourea." Translated from Izv. Akad. Nauk. SSR, Met., Vol. 6, 1975, pp. 32-7.

17. Chtyan, G., Vartanyants, S., Adzhemyan, O., Andresyan, D., and Babayan, G., "Possibility of extracting gold using an acid solution of thiourea in the presence of ozone. II." Arm. Khim. Zh., Vol. 29, No. 32, 1976, pp. 225-9 (Russ).

18. Chtyan, G. and Babayan, G., "Recovery of gold and silver in an acid thio-carbamide solution under the action of ozone," Mater.-Vses. Meshvus. Konf. Ozonn, 2nd, 1977, pp. 17-8 (Russ).

19. Muldagalieva, I., Rozhdestvenskaya, Z., and Songina, O., "Study of the anodic dissolution of gold and silver in the presence of oxidants and complexing agents," Tezisy Dokl.-Vses. Soveshch. Khim., Anal. Tedhnol. Blagorodn. Met., 10th, Vol. 1, 1976, p. 103 (Russ).

20. Muldagalieva, I., Rozhdestvenskaya, Z., and Songina, O., "Effect of oxidants on anodic behavior of copper subgroup metals," Zh. Anal. Khim., Vol. 3, No. 1, 1976, pp. 176-9 (Russ).

21. Tartaru, S., "Dissolution of gold in thiourea solution," Rev. Roum. Chim., Vol. 13, No. 7, 1968, pp. 891-7 (Fr).

22. Lodeishehikov, V., Panchenko, A., and Bryantseva, L., "Use of thiourea as a solvent for gold," Nauch. Tr., Irkutsk. Gos. Nauch.-Issled. Inst. Redk. Tsvet. Metal., Vol. 19, 1968, pp. 72-84 (Russ).

23. Lodeishchikov, V. and Panchenko, A., "Evaluation of gold solvents," Obogasch. Rud. Peskov Blagorod. Metal., 1971, pp. 122-4 (Russ).

24. Songina, O., Ospanov, Kh., Muldagalieva, I., and Sal'nikova, S., "Dissolution of gold with the use of thiourea in a hydrochloric acid medium," Izv. Akad. Nauk. Kaz. SSR, Ser. Khim., VCol. 21, No. 3, 1971, pp. 9-11 (Russ).

25. Klimenko, N., Savori, E., Koroleva, E., and Shvedovo, I., "Dissolution of gold by acidic thiourea solutions," Tr., Tsentr. Nanchno-Issled. Geologo-razved. Inst. Tsvetn. Blagorodn. Metal., Vol. 111, 1974, pp. 57-61 (Russ).

26. Pirmagomedov, D., Khalezov, B., Kakovskii, Brik, I., and Abdulova, A., "Processing of polymetallic ores with a thiourea solution for extracting noble metals," Obogashchenie Rud (Irkutsk), Vol. 5, 1977, pp. 165-71 (Russ).

27. Nagy, i., Mrkusic, P., and McCulloch, H. W., "Chemical treatment of refractory gold ores," Literature survey, Republic of South Africa, National Institute for Metallurgy, Project C, 28/62, Report No. 38, June 1966, pp. 4-11.

28. Henley, K. J., "Gold-ore mineralogy and its relation to metallurgical treatment," Minerals Sci. Engng., Vol. 7, No. 4, Oct. 1974, pp. 303-5.

29. Wargo, J. G., "The next exploration stage for Carlin-type gold deposits," Mining Engineering, Vol. 31, No. 9, 1979, pp. 1321-3.

30. Radtke, A. and Dickson, F., "Genesis and vertical position of fine-grained disseminated replacement-type gold deposits in Nevada and Utah, USA." Internat. Assoc. Genesis Ore Deposits (IAGOD) Proc., 4th Symposium, Varna, Bulgaria, Vol. 1, 1974, pp. 71-8.

31. Merwin, R. W., "Gold resources in the oxidized ores and carbonaceous material in the sedimentary beds of northeastern Nevada," Washington, USBM, Technical Progress Report 1, March 1968, 16 pp.

32. Bugbee, E. E., A Textbook of Fire Assaying, New York, John Wiley and Sons, 1933, 314 pp.

33. Yatsimirskii, K. B. and Astasheva, A. A., "The use of thiourea solutions in volumetric analysis (thiocarbamidometry)," Journal of Analytical Chemistry of the USSR, Vol. 11, 1956, pp. 465-6.

34. Groenewald, T., "The leaching of gold and uranium in acidic solution of thiourea," Chamber of Mines of South Africa Research Organization, Project No. 11/504/54, Progress Report No. 1, August 1970, 16 pp.

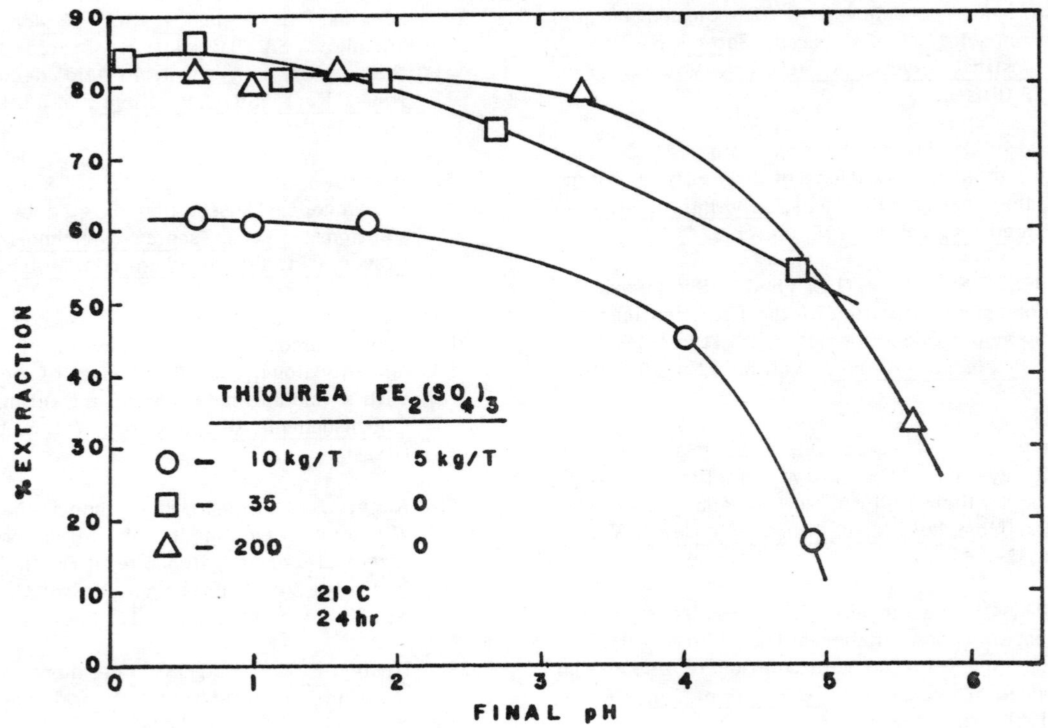

FIG. 1. PERCENT GOLD EXTRACTION VS. pH FOR CHIPMUNK ORE

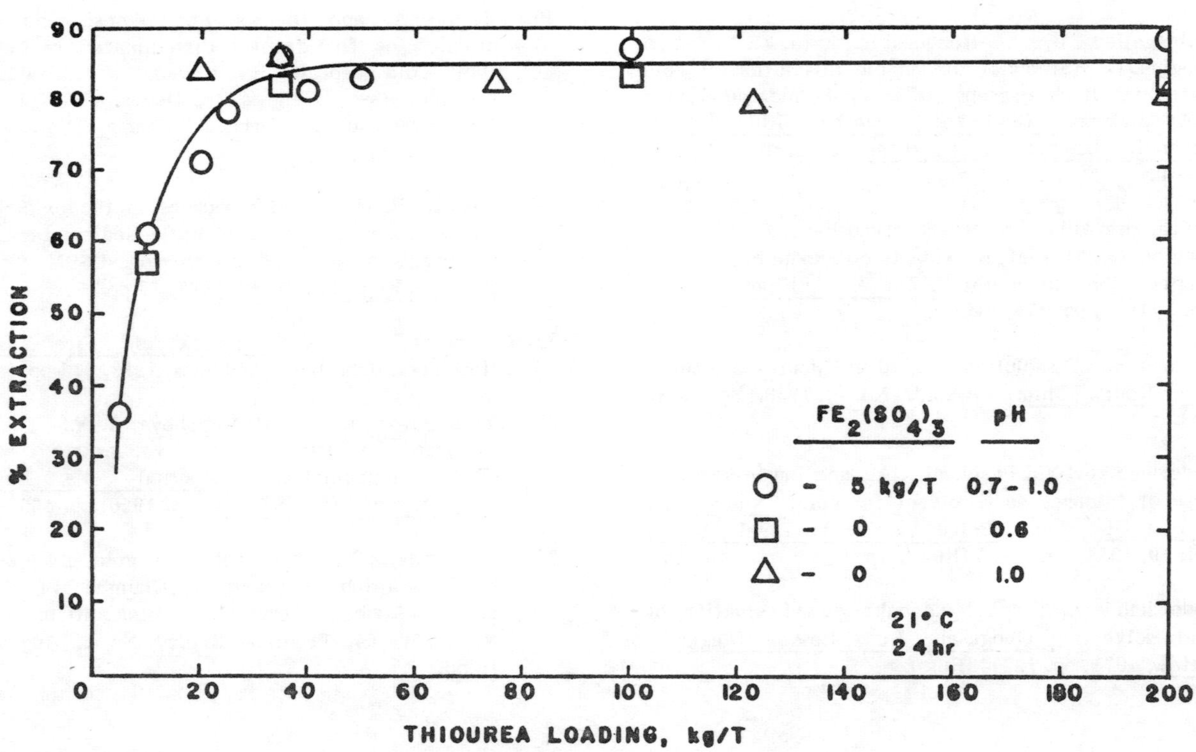

FIG. 2. PERCENT GOLD EXTRACTION VS. THIOUREA LOADING FOR CHIPMUNK ORE

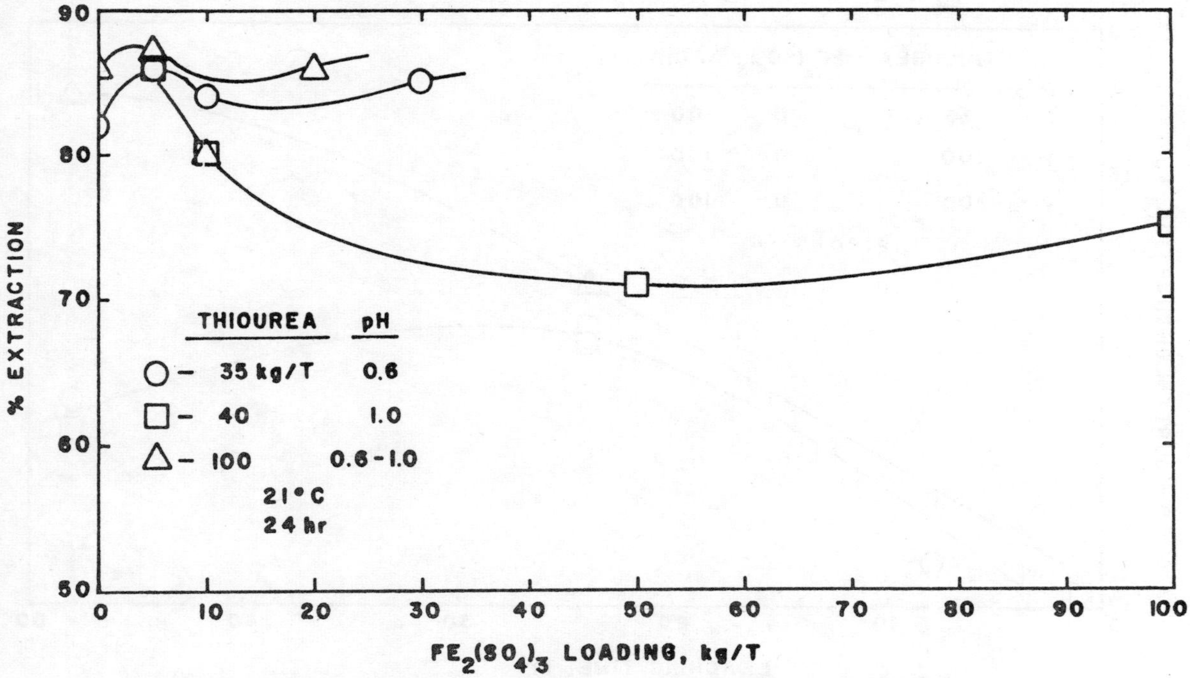

FIG. 3. PERCENT GOLD EXTRACTION VS. $FE_2(SO_4)_3$ LOADING FOR CHIPMUNK ORE

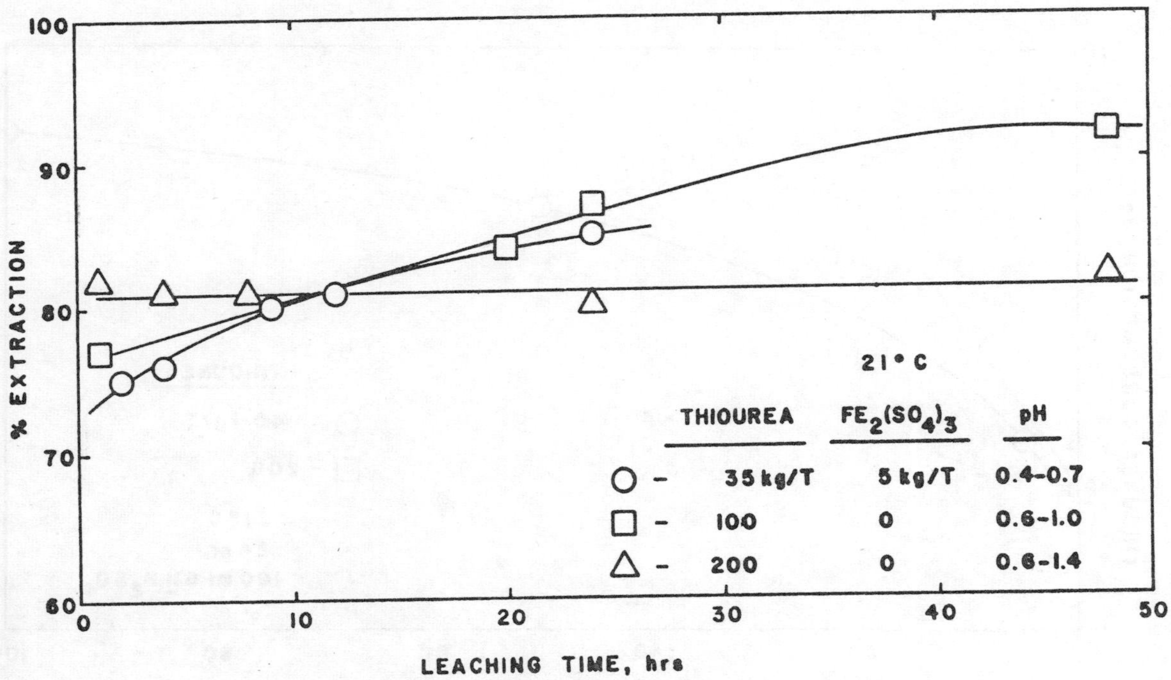

FIG. 4. PERCENT GOLD EXTRACTION VS. LEACHING TIME FOR CHIPMUNK ORE

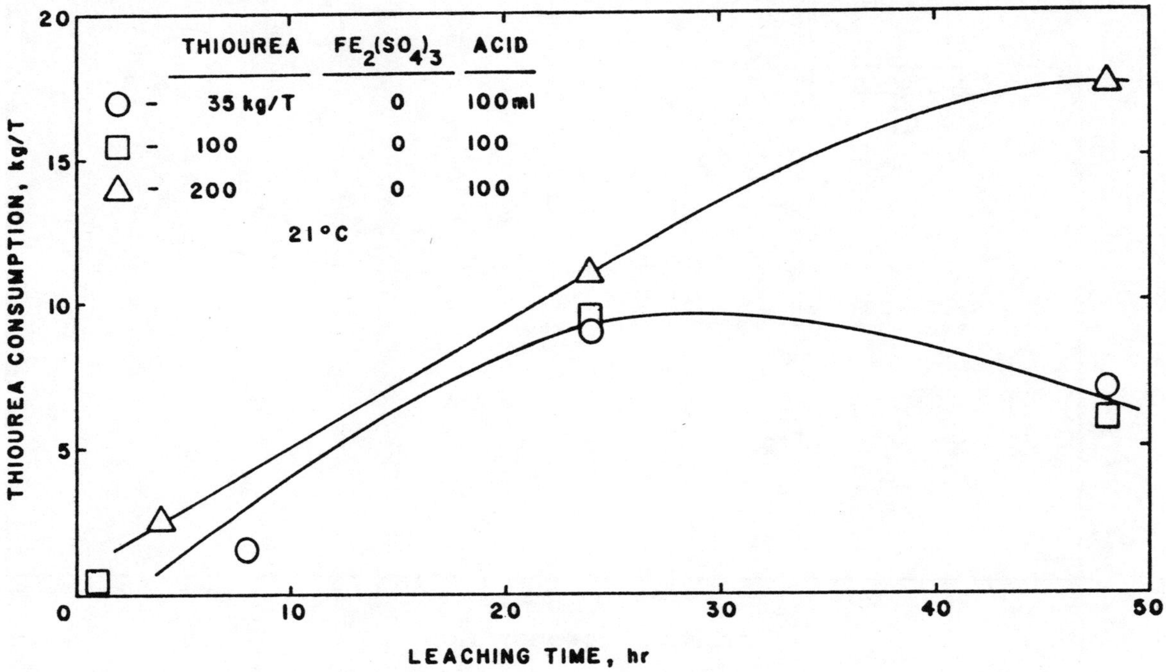

FIG. 5. THIOUREA CONSUMPTION VS. LEACHING TIME FOR CHIPMUNK ORE

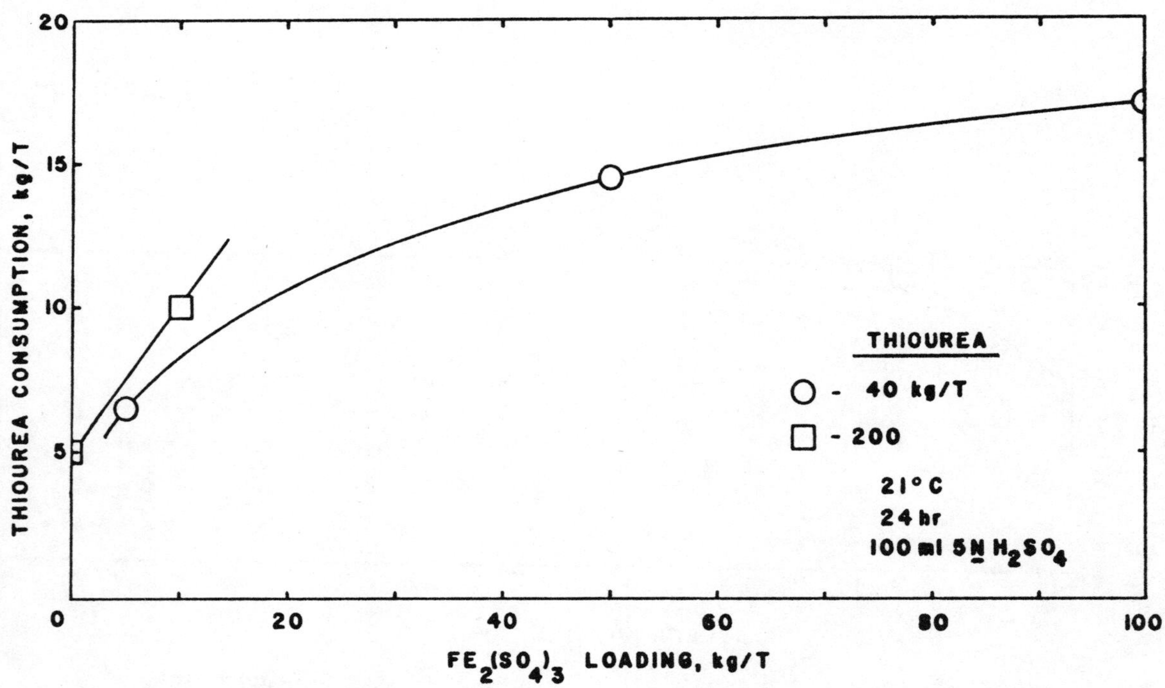

FIG. 6. THIOUREA CONSUMPTION VS. $FE_2(SO_4)_3$ LOADING FOR CHIPMUNK ORE

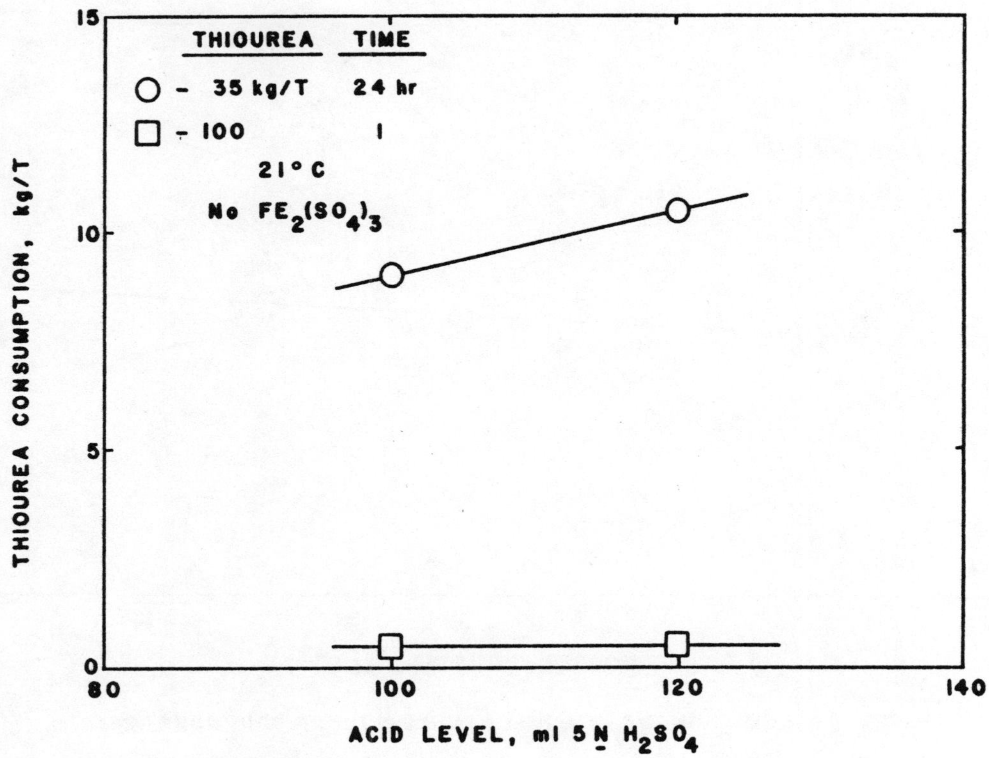

FIG. 7. THIOUREA CONSUMPTION VS. ACID LEVEL FOR CHIPMUNK ORE

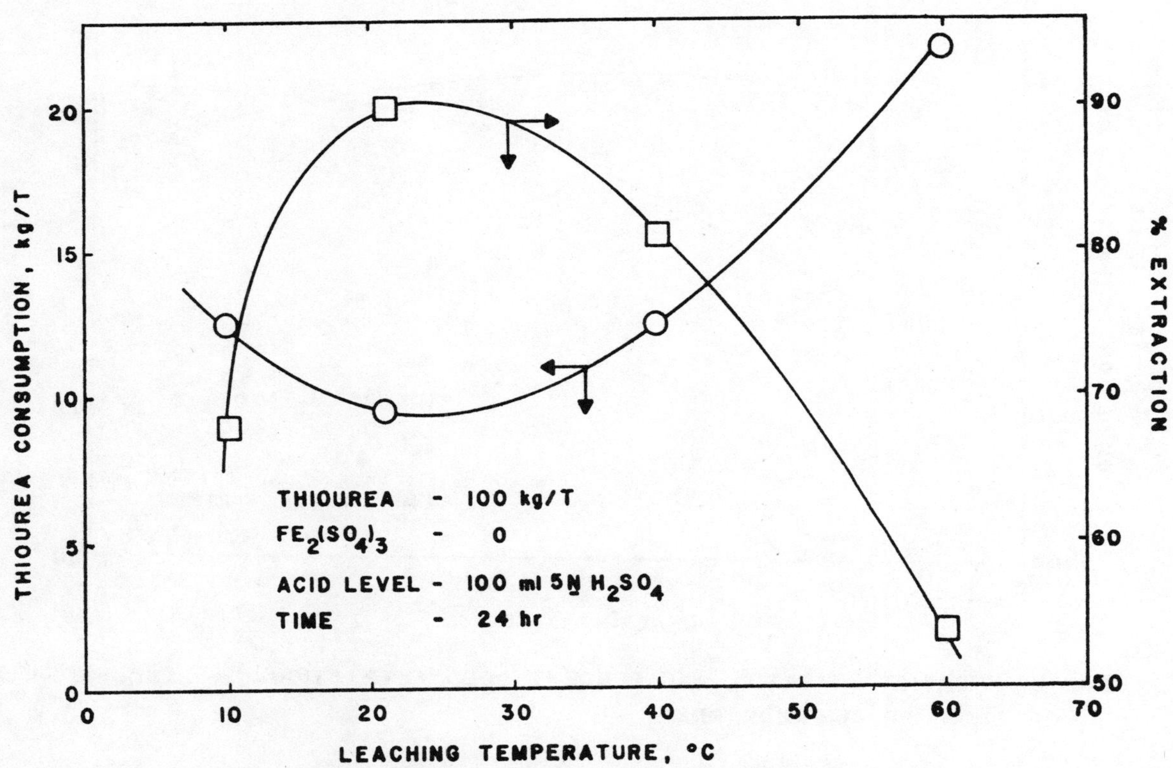

FIG. 8. THIOUREA CONSUMPTION AND PERCENT EXTRACTION VS. LEACHING TEMPERATURE FOR CHIPMUNK ORE

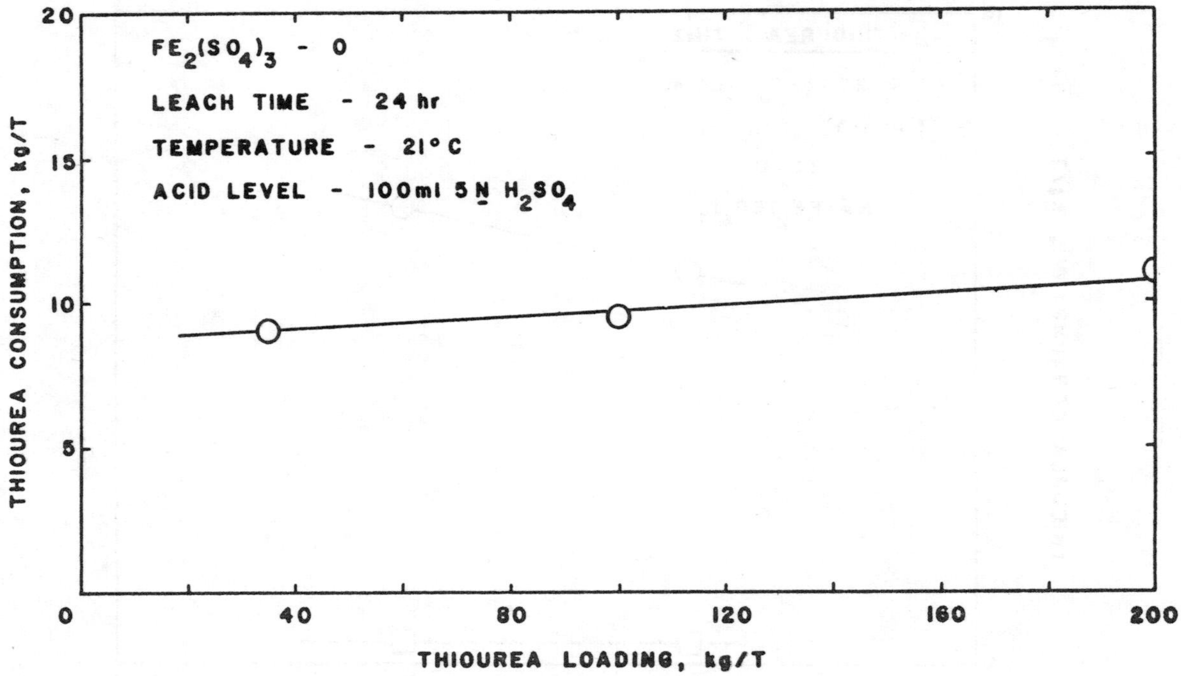

FIG. 9. THIOUREA CONSUMPTION VS. THIOUREA LOADING FOR CHIPMUNK ORE

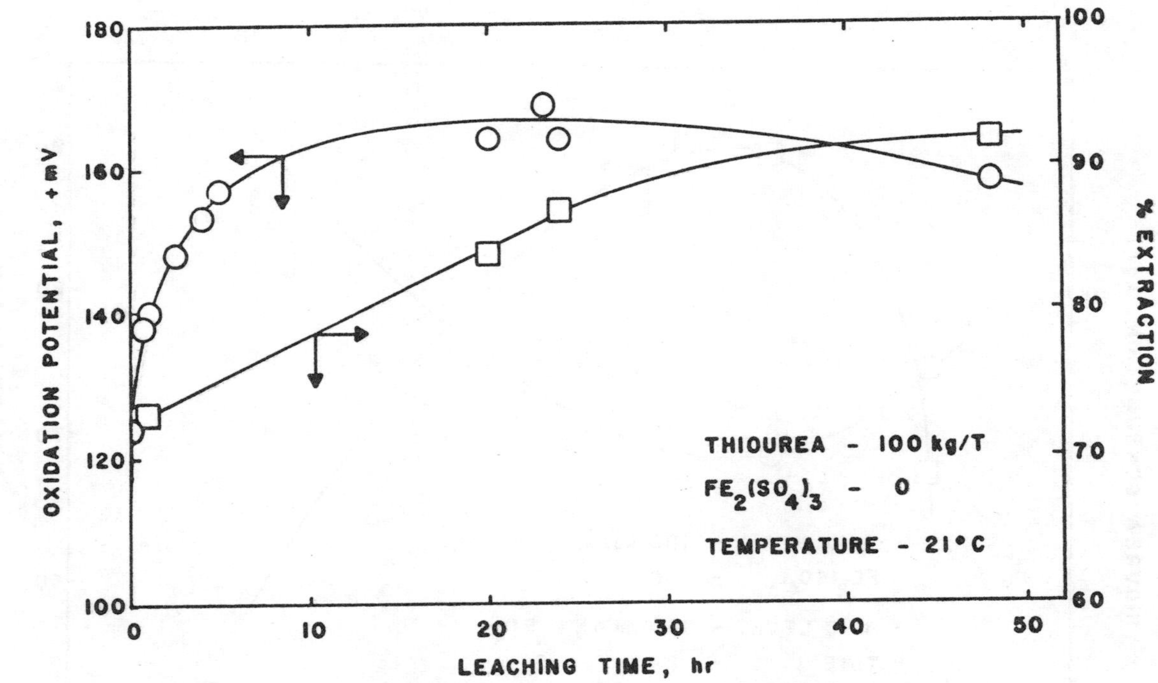

FIG. 10. OXIDATION POTENTIAL AND PERCENT GOLD EXTRACTION VS. LEACHING TIME FOR CHIPMUNK ORE

Chapter 12

GOLD AND SILVER ASSAY METHODS IN THE MINING AND METALLURGY INDUSTRY

T. N. Anderson
J. C. Parr
L. C. Stone
D. Metcalf
Process Technology
Kennecott Minerals Company
Salt Lake City, Utah 84147

ABSTRACT

An increase in the importance of low-grade materials as sources of gold and silver has led to the need for improved analytical methods on leach solutions, ores and various effluents, residues, and in-process streams. These are required by plant operators as guides to process control and for accounting purposes. Atomic absorption spectrophotometry (AAS) and AAS preceded by pre-concentration have been applied successfully to resolve many such problems. However, AAS is no panacea and imprecision or loss of accuracy can result if adequate precautions are not taken to prevent losses and eliminate interferences. The recent methods are discussed together with methods for overcoming difficulties. Comparisons with more classical analytical techniques are also given.

Introduction

The history of gold and silver analyses shows an interesting duality. On one hand, classical chemical and instrumental methods have evolved to meet changing needs, but on the other hand, the age-old and somewhat empirical fire assay has remained a popular tool. Inasmuch as gold and silver generally have been present in ores at low concentrations, methods have been challenged by a need for sensitivity and selectivity coupled with reasonable assay time. A progressive increase in the importance of low grade materials as ore sources has led to the need for improved methods. Assays of these materials are required for process control and accounting in operating plants as well as for guiding research and development activities. The analytical response to these needs is evidenced by numerous reviews published through 1977--e.g., cf Kolthoff and Elving (1966), Beamish and Van Loon (1966, 1972, and 1977), Young (1971), Mallett (1970), and Sen Gupta (1973).

In this paper we present a brief sketch of the history and status of various gold and silver methods. A more detailed summary of atomic absorption spectroscopy (AAS) is given, as this method has gained in popularity more than other assay methods. AAS methods currently in use are summarized and discussed briefly. This paper is not meant to be a detailed review, as are the above mentioned publications. Rather, we borrow from those works and more recent papers to report trends in gold and silver analyses (particularly in AAS). Also, we discuss some selected potential problems, and problems that have been experienced by this laboratory.

Status of Au and Ag Methods

Table 1 summarizes the status of the more important gold and silver methods. All of these have found use in the industry at some time, and most are probably used occasionally by those laboratories which are set up for ready application. However, fire assay and AAS account for most quantitative work at present.

Fire assay has been the time-honored method for gold and silver determinations. Applied properly, it combines a very efficient metal collection or preconcentration technique with the primary detection method of weighing the gold bead. Other methods can be used to ascertain its purity. Accuracy for gold assays is usually acknowledged to be high, even though the claim is made occasionally that an ore contains values which fire assay has missed. (After forty years of research in the field, Beamish (1972) claimed that he had not experienced a single example of failure of the classical assays to find a paying ore!). By combining lead buttons from various sample collections, very large samples can be analyzed with acceptable precision and good representation of heterogeneous materials. The vitality of fire assay is attested to by the fact that it is still the most used method for umpire and accounting purposes. Also, a book has been published very recently by Smith (1979) which details procedures for fire assay.

Additional and newer methods have been and are important to laboratories which have applicable instrumentation, which have no fire assay equipment or analyst, or which find fire assay cumbersome--e.g., as in the case of liquids, few samples per day, or very low grade solid samples. These methods also have contributed significantly to the general state of analytical quality control, by proving the accuracy of the more routine methods with changing sample type. Thus, most laboratories at gold and silver mines use atomic absorption as well as fire assay.

Unless the fire assay losses are determined (which is quite lengthy), silver assays are more accurately made by other means. These cupel losses, which are of the order of several percent for typical ores (see references in Mallett [1970] and Walton [1973]), have been accepted possibly because toll smelters expect to lose a comparable fraction of the silver during processing. With hydrometallurgical processes, toleration of these errors cannot be justified.

Fire assay is not sensitive enough for the practical assaying of many low grade ores or of small samples of nominally low grade material, which result from many laboratory or exploration studies. In these cases, fire assay collection is often combined with the more sensitive detection methods of atomic absorption or emission spectrography.

Before the development of AAS, most of the methods in Table 1 found some mining or metallurgical application. Gravimetric and titrimetric methods were used when richer ores were considered. They still find some use for higher grade intermediate and end metallurgical products when good precision is needed (e.g., copper anode sludges). Spectrophotometric (i.e., colorimetric) methods were very popular for trace levels of Au and Ag in the 1950's and 1960's. Electrometric methods also are very sensitive and are also very precise, but require more sophisticated instrumentation, so are used mostly where equipment is available and used for other purposes. X-ray methods and emission spectrography are useful, where applicable, because of their multi-element capabilities and accommodation of solid samples. Both suffer from variable matrix interference and, therefore, are quantitative only if complemented by adequate standards. Emission spectrography has been used as a quantitative method both for fire assay products and for noble metals which were separated from the base metals by precipitation. It continues to be used semiquantitatively for trace levels of noble metals in many materials, and for quantitative assays in relatively simple matrices for which enough reliable standards are available (e.g., in copper). Neftron activation is uniquely applicable to small (e.g., 1 g) samples which contain ultratrace levels of Ag or Au. As such, it appears to have more application in geologic than mining and metallurgical applications.

Atomic absorption started to find application in the early and middle 1960's. Since that time it has gained wide usage for all types of mining and metallurgy samples. Today it is the mainstay method for gold and silver in many or most laboratories and is important with regard to these assays in nearly all laboratories.

Some laboratories use rapid but probably incomplete sample digestions coupled with AAS procedures to yield quick, semiquantitative results. However, with the appropriate sample and standard preparation, AAS regularly gives gold results rivaling fire assay accuracy, and silver results which exceed it, considering the slag and cupel losses in fire assay (e.g., see Mallett [1970], Tindall [1965], and Fishkova, et al. [1972]). Although fire assay is heralded for its greater precision than other methods, proper sample preparation coupled with AAS yields equal or better precision on typical ores. This was demonstrated by Brandvold (1979) and in other recent literature. Brandvold's tests involved eight commercial laboratories which used fire assay and seven analysts (from commercial, industrial, and government laboratories) which used AAS. Both interlaboratory and intralaboratory results, for gold and silver (at 0.25 oz/ton and 17 oz/ton, respectively) were at least as precise for AAS as for fire assay. At much higher ore grades, it is recognized that fire assay has the advantage over AAS. Tests conducted by our laboratory confirmed this conclusion in assays of copper refinery slimes, which contained gold and silver at levels several orders of magnitude higher than those in ores. The fire assays were up to several hundred percent more precise than AAS in this test.

Au and Ag Atomic Absorption Methods

General

AAS gained in popularity over other nonfire methods because of its low cost, versatility, simplicity of operation, and specificity. Many aqueous samples can be aspirated into the flame with no pretreatment or with just a water dilution, and many solids require little more than a simple mixed acid digestion. Matrix effects from base metals and the solvents are usually much less than in the case of other instrumental methods. Modifications to the direct AAS procedures came about largely for low grade materials in which the gold or silver signal was nearly the same as background. Table 2 gives an overview of atomic absorption methods for the most frequently encountered sample types, over a wide range of silver and gold concentrations. Table 3 details some solvent extraction methods.

Discussion of Methods and Potential Problems

Sample Preparation: Aqua regia with the addition of hydrofluoric acid, when needed, is the most-used lixiviant, and is applicable to most ore types. Other combinations of acids as well as cyanides are also used, and preroasting is sometimes applied.

Digestion time and amount of reagent vary, quite naturally, with the amount of acid-consuming materials in the sample. Thus, raw ores are usually simpler than base metal and pyrite concentrates. In the latter case, reagents in addition to aqua regia are often added to completely oxidize the sulfide; otherwise, sulfur beads may form on the reaction mixture.

Detailed digestion procedures vary with the sample type, size of gold particulate, and laboratory. No one precedure or even reagent combination is optimum for all types of samples. This was demonstrated within our laboratory, where aqua regia is the standard lixiviant for most solid samples. However, silver in a carbon bed (used to remove the Ag from cyanide leach liquors) could not be dissolved quantitatively with either an aqua regia or a cyanide leach. It was finally dissolved quantitatively after the carbon was completely destroyed with hot, concentrated, $HClO_4$.

After digestion, the sample is redissolved in the solvent from which AAS measurements or solvent extraction is to take place. The sample is usually boiled or baked before the final solution makeup. General consensus is that this is done to remove NO_x. The reasons for this are not altogether clear, but it has been reported that NO_x interferes with AAS silver readings (Tindall, 1977), and some gold methods also have included baking. However, methods for both Au and Ag are used which do not include such baking, and the excess nitric acid does not interfere with subsequent extraction of gold into MIBK (Ichinose, 1971). Therefore, the general need for baking appears to be an unresolved question at present.

Independent of the above question, it has been reported that boiling samples to dryness and then baking at high temperatures can reduce Au(III) to Au(I) and Au, which do not extract into ketones such as MIBK (Tindall [1977]). Therefore, Tindall recommends that the sample be kept moist and NO_x be removed by boiling from, e.g., 50% HCl. Some authors add oxidants such as bromine or permanganate to the final makeup solution to counteract any prior reduction of Au(III).

Both AAS measurements and extraction can be performed from HCl or aqua regia solutions, the same components used in digestion. Although Au (III) is sufficiently soluble in even dilute HCl, silver may be lost to precipitation if the HCl is not concentrated enough. Figure 1 gives the solubility of Ag+ ions as a function of HCl concentration. This figure does not take into consideration chloride used in complexing other metal ions such as Fe^{3+}, Al^{3+}, and Pb^{2+}. Common practice is to make the final dissolution in 25 to 50% HCl which, according to the literature, is sufficient to hold at least 50 ppm Ag in solution. Ammonium acetate (NH_4Ac) is often added to the HCl solution when silver is determined in the presence of high lead. Although this improves the solubility of silver slightly (cf. Fig. 1) and prevents $PbCl_2$ precipitation (which can coprecipitate silver, cf Tindall [1965]), the value of ammonium acetate is not universally accepted (see Tindall [1966]).

Leach solutions such as cyanides may be aspirated directly into the flame. Standards should match the samples quite closely both in the case of solution samples and digest-makeup samples. Reliable background correction is essential for quantitative work, especially in the case of gold, unless both the standards and sample matrices are nearly identical. This is difficult to accomplish inasmuch as different samples usually contain different, unknown levels of the base metals and solvents. Background correction is practically a necessity with flameless AAS.

Air-acetylene flames and wavelengths of 328.1 nm (for Ag) and 242.3 nm (for Au) are employed almost universally. Typical sensitivities* are approximately 0.05 ppm for Ag and 0.3 ppm for Au.

Solvent Extraction: As Au and Ag concentrations decrease and additional sensitivity is needed, solvent extraction (SX) is usually employed. This serves to concentrate the sample, remove interferences (discussed later), and enhance the AAS signal for a combined sensitivity increase of up to 200 times. A suitable organic solvent provides quantitative extraction from the digestate or solution sample at a small organic/aqueous ratio, low solubility in the aqueous phase, selective extraction of Au or Ag over the other ions present, a flame transparent at the wavelengths involved, and enhancement of the absorption. The better and more popular solvents in these respects are listed in Table 3. Many others have been studied, however, as evidenced by the more than 100 SX schemes for gold and more than 20 for silver which were published from 1970-80.** Most of these references can be found readily in the reviews by Sen Gupta (1973) and Das and Bhattacharyya (1976), and in the biannual bibliographies of Atomic Absorption Newsletter. The great differences in these numbers is no doubt due to the greater sensitivity of the AAS response for silver, coupled with the fact that commercially interesting silver deposits are of higher grade than their gold counterparts.

Most gold extraction is done from HCl or aqua regia solutions containing chloride, 10 to 60%, but a few extractants function in cyanide or both cyanide and acid solutions. Silver may be extracted from similar solutions, but often requires additional treatment. Although the TOPO method of Burke (1974) was originally developed for Ag in metallic nickel, the authors have successfully applied it to ores and other mining materials. Cyanide leachates have been converted to HCl systems, preceded by oxidation of the Au(I) to Au(III) with such oxidants as $KMnO_4$, aqua regia, and Br_2.

Although pre-equilibration of the extraction solution and the aqueous acidic media is deemed advisable by some authors, it has not been practiced universally (e.g., cf Tindall [1965]). Pre-equilibration, however, ensures constant recovery of the organic phase, when the solvent system solubility is not negligible--e.g., MIBK is soluble to the extent of 2 ml per 100 ml in 0.6 M HCl and approximately 4 ml per 100 ml in 6 M HCl.

*The sensitivity is usually defined as the concentration which produces a 1% absorption.

**Not all of these schemes are coupled with AAS detection.

Ion Exchange: Resins also have been used for purifying and concentrating Au and Ag solutions prior to AAS analysis. A review of silver work is given by Kolthoff and Elving (1966) and one for gold work is given by Beamish and Van Loon (1977). This laboratory has recently used Rohm and Haas' XAD-7 to concentrate gold from mining samples, dissolved in aqua regia and brought up in HCl solutions. The gold was then stripped with an acetone-ethanol mixture or 40% HNO_3 and aspirated into the flame.

Flameless AAS

Flameless AAS provdes a means to increase flame AAS sensitivity by up to 200 times. Interferences from aqueous leachates and digestates are so great, however, that flameless AAS has found application mainly after organic extraction. Thereby, the analytical sensitivity is increased far beyond that of the SX/flame-AAS methods.

Interferences: Interferences from other elements have been reported as minimal in many cases, but matrix difficulties and precautionary steps to avoid them are also prevalent in the literature (see Mallett [1970] and Sen Gupta, [1973]). High salt content, from some elements more than others, causes AAS interferences from broad band spectra and light scattering--i.e., Fe^{3+}, NaCN, KCN, Zn, Se, Te, Ni, Co, and Cu in the case of gold and Fe^{3+}, NaCN and KCN in the case of silver. Unless the salts are very concentrated, e.g., as when several grams of soluble sample are dissolved in 100 ml, we have found that background correction eliminates such interferences. Without background correction, we have found that quantitative results are difficult to attain. As reliable background correctors were not common until the late 1960's, it may be assumed that many problems reported were, in fact, due to the lack of this important AAS accessory. Even when a background corrector is used, we strongly recommend that its functionality be tested frequently with a matrix blank which matches the sample. This sould read zero if the background corrector is working.

At very high salt concentrations, the background corrector often cannot do its job; the high viscosity of concentrated solutions also impedes aspiration. In the extreme, salts may collect on the burner head. Although complexing agents have sometimes been added to increase solubilities of some species, solvent extraction appears the more common remedy. Even then, some species, e.g., Fe, may extract into the organic, rendering its properties unsuitable for aspiration. For this reason, the MIBK is given several washes with dilute (e.g., 1.5 M) HCl to remove interference, the gold remaining in the organic phase.

Determination of Gold and Silver from a Single Sample Preparation: Gold and silver are often determined from the same sample, so a single sample preparation prior to AAS measurement offers substantial time savings. Single sample preparation is, of course, easily accomplished when both metals dissolve in the digest (e.g., aqua regia) or the leach solution (e.g., cyanide) at levels measurable with flame AAS.

Often the gold/silver ratio dictates that gold must be extracted but that silver can be determined directly from the digest. Tindall (1965) accomplished this from an aqua regia digest with 25% HCl makeup, and solvent extraction of gold into MIBK. Silver was determined from the HCl before the gold extraction (this is necessary as silver extracts incompletely into the MIBK), and gold was determined from the MIBK. The gold standards were prepared by extraction into the ketone from 25% HCl.

When this procedure was attempted by the authors with 5 g copper concentrates (which contained 25-30% Fe) and 150 ml of 4.8 M HCl as makeup, it was found that the gold was not extracted completely into 10 ml of MIBK. This appeared to result from overloading the MIBK with iron. At dilute HCl concentrations (e.g. 1.8 M), the gold was extracted completely but the silver was not stable in the aqueous phase. This exemplifies the difficulty in developing methods which are applicable to all sample types and indicates the need for more work in methods development for the determination of gold and silver from a common digestion.

It would be advantageous to extract both the gold and silver from a common lixiviant (e.g., aqua regia), and aspirate them from a common organic solvent. This has been done (see Konovalov and Gureva, 1979) for high levels of Au and Ag in Sb-Pb alloys using petroleum sulfides in toluene as the extractant. Considering flame characteristics of this solvent, it would appear more suitable for ores to use some of the extractants in Table 3 which are common to both silver and gold--e.g., TOPO in MIBK or Aliquat 336 in DIBK. Such a method could not be found in the literature. It also appears possible to dissolve these metals, after aqua regia digestion and baking, with a NaCN solution and to either extract both metals in Aliquat 336/DIBK from that solution or directly aspirate the silver and extract the gold.

RECENT TRENDS IN SILVER AND GOLD METHODS

Literature of the last several years shows that strong effort is continuing in expanding, improving, optimizing, and evaluating analytical methods for gold and silver in mining and metallurgical applications. Well over 100 articles have been published since 1973 dealing just with atomic spectroscopy. A complete listing is beyond the scope of this article; again the reader is referred to the biannual bibliographies of <u>At. Absorpt. Newsl.</u> and <u>Atomic Spectroscopy</u>.

The accuracy, sensitivity and precision of AAS, applied properly, are at least as good as those for fire assay. Therefore, the choice between these methods would appear mainly to be one of assay time and available analyst skills. Flameless AAS is receiving more attention, and the result is nanogram or picogram sensitivities. Most separation techniques are still being considered--i.e., conventional ones such as solvent extraction, precipitation and fire assay, and also recent ones (e.g., see Kahn and Van Loon, 1978, who used ion exchange chromatography).

The state of present methods probably leaves assay time as a more important problem in the mining industry than sensitivity or method availability, per se. Therefore, innovations to minimize the number of assay steps and to automate are being pursued (see Hall [1979], Parkes and Murray-Smith [1979], and Brandt, Van Dalen, and Wessels [1980]. From 50 to 100 samples can be run in one man shift (8 hr) using present technology.

In a more exploratory vein, both flame and flameless AAS have been attempted directly on solid samples. In the flame AAS case, see Karmenova and Pogrebuysk, 1979, the sample was diluted with powdered graphite and this mixture aspirated into the flame.

Multielement analysis from both the raw sample, digests, and digest/extracts of the sample are probably the most considered time-saving procedure. Emission spectrography is far from dead, and several recent publications have applied it to the raw samples, so as to omit time-consuming separations. As discussed previously, single solvent extractions and also precipitations are developed to separate and concentrate several elements simultaneously. Publications applying this procedure are most prevalent for gold in conjunction with the platinum metals.

Perhaps the greatest time saving innovation developed during the '70's will be inductively coupled plasma (ICP) spectroscopy. Although applications to gold and silver are still sparse (e.g., see Motooka, et al., 1979, and Wemyss, et al., 1978), increased use is expected. ICP offers linear response over a much wider concentration range than AAS and, with the higher temperature, offers added spectral lines and sensitivity. Therefore, it is conceivable that samples soon may be determined by reading them with ICP after a simple digestion, and later the solids may be determined directly. With such capabilities the high cost of ICP equipment could be justified for many laboratories.

References

1. Aruscavage, P. J., and Campbell, E. Y., (1979), "Determination of Silver in Silicate Rocks by Electrothermal Atomic Absorption Spectrometry," <u>Anal. Chim. Acta.</u>, Vol. 109, p. 171.

2. Beamish, F. E., 1966, <u>The Analytical Chemistry of the Noble Metals</u>, Pergamon Press, Oxford, England.

3. Beamish, F. E. and VanLoon, J. C., 1972, <u>Recent Advances in the Analytical Chemistry of the Noble Metals</u>, Pergamon Press, New York.

4. Beamish, F. E., and VanLoon, J. C., 1977, <u>Anlaysis of Noble Metals</u>, Academic Press, New York.

5. Brandt, P. J., Van Dalen, J. H., and Wessells, F. W.,1980, "The Automatic

Determination of Trace Amounts of Gold in Gold-Plant Barren Solutions," J. So. Afr. Inst. Min. Metall., June 1980, p. 197.

6. Brandvolt, L. A., 1979, "Reliability of Gold and Silver Analyses by Commercial Laboratories in the Southwest, "New Mexico Geology, February, 1979, p. 11.

7. Bratzel, M. P., Jr., Chakrabarti, C. L., Sturgeon, R. E., McIntyre, M. W., and Agemian, H., 1972, "Determination of Gold and silver in Parts-per-billion or lower levels in Geological and Metallurgical Samples by Atomic Absorption Spectrometry with a Carbon Rod Atomizer," Anal. Chem., Vol. 44, p. 372.

8. Burke, K. E., 1974, "Determination of Microgram Quantities of Silver in Aluminum-, Iron-, and Nickel-Base Alloys," Talanta Vol. 21, p. 417.

9. Das, N. R. and Bhattacharyya, S. N., 1976, "Solvent Extraction of Gold," Talanta, Vol. 23, p. 535.

10. Fishkova, N. L., Fal'kova, O. B., and Meshalkina, R. D., 1972, "Atomic Absorption Determination of Gold in Ores and Evaluation of the Effect of the Inhomogeneity of its Distribution in Samples on the Analysis Results," Zh. Anal. Khim., Vol. 27, No. 10, p. 1916.

11. Greaves, M. C., 1963, "Determination of Gold and of Silver in Solution by Atomic Absorption Spectroscopy, Nature, Vol. 199, p. 552.

12. Groenewalt, T., 1968, "Determination of Gold(I) in Cyanide Solutions by Solvent Extraction and Atomic Absorption Spectrometry," Anal. Chem., Vol. 40, No. 6, p. 863.

13. Groenewald, T., 1969, "Quantitative Determination of Gold in Solution by Solvent Extraction and Atomic Absorption Spectrometry," Anal. Chem., Vol. 41, p. 1012.

14. Hall, S. H., 1979, "A Rapid Method for Gold Extraction using MIBK," At. Absorpt. Newsl., Vol. 18, No. 6, pp. 126-7.

15. Ichinose, N., 1971, "Extraction and Determination of Metal Salts with Isobutyl Methyl Ketone. XIV. Extraction of Gold," Talanta, Vol. 18, No. 1, p. 105.

16. Kahn, N. and VanLoon, L. C., 1978, "Direct Atomic Absorption Spectrophotometric Analysis of Anion Complexes of Platinum and Gold after Concentration and Separation from Aqueous Solutions by Anion Exchange Chromatography," Anal. Lett., Vol. All, p. 991.

17. Karmanova, N. G. and Pogrebnyak, Yu. F., 1979, "Atomic-Absorption Determination of Silver in Sulfide Ores," Zavod. Lab, Vol. 45, No. 2, p. 124. see CA 90:197069s.

18. Kolthoff, I. M. and Elving, P. J., 1966, Treatise on Analytical Chemistry, Part 2 Vol. 4, Interscience Publishers, New York, N.Y., p. 1.

19. Konovalov, G. F. and Gur'eva, M. P., 1979, Extraction-Atomic Absorption Determination of Silver and Gold," Zavod. Lab, Vol. 45, No. 3, p. 196 (see CA 90: 214674e).

20. Mallett, R. C., 1970, "A Review of Techniques for the Determination of Gold and Silver by Atomic-Absorption Spectroscopy," Miner. Sci. Engng., Vol. 2, p. 28.

21. Mallett, R. C., Taylor, J. D., and Steele, T. W., 1966, "The Determination of Gold in Barren Cyanide Solutions by Atomic Absorption Spectrophotometry," Johannesburg, National Institute for Metallurgy, Research Report No. 24.

22. Motooka, J. M., Mosier, E. L., Sutley, S. J., and Viets, J. G., 1979, "Induction-Coupled Plasma Determination of Silver, Gold, Bismuth, Cadmium, Copper, Lead, and Zinc in Geologic Materials using a Selective Extraction Technique - Preliminary Investigation," Appl. Spectrosc., Vol. 33, p. 456.

23. Parkes, A. and Murray-Smith, R., 1979, "A Rapid Method for the Determination of Gold and Palladium in Soils and Rocks," At. Absorpt. Newsl., Vol. 18, p. 57.

24. Purushottam, A., Lal, S. S., and Naidu, P. P., 1972, "Rapid Determination of Traces of Silver in Sulphide Ores by Atomic Absorption," Talanta, Vol. 19, p. 208.

25. Rubeska, I., Koreckova, J., and Weiss, D., 1977, "The Determination of Gold and Palladium in Geological Materials by Atomic Absorption after Extraction with Dibutyl Sulfide," At. Absorpt. Newsl., Vol. 16, p. 1.

26. Sen Gupta, J. G., 1973, "A Review of the Methods for the Determination of the Platinum-Group Metals, Silver, and Gold by Atomic-Absorption Spectroscopy," Miner. Sci. Engng., Vol. 5, p. 207.

27. Smith, S. L., 1979, Fire Assaying for Gold and Silver, Jacobs Assay Office, Tucson, Arizona.

28. Sukiman, S., 1976, "Determination of Gold in Ores by Atomic Absorption Spectrometry after Chromatographic Separation," Anal. Chim. Acta, Vol. 84, p. 419.

29. Tindall, F. M., 1965, "Silver and Gold Assay by Atomic Absorption Spectrophotometry," At. Absorpt. Newsl., Vol. 4, No. 9, p. 339.

30. Tindall, F. M., 1966, "Notes on Silver and Gold Assay by Atomic Absorption," At. Absorpt. Newsl., Vol. 5, No. 6, p. 140.

31. Tindall, F. M., 1977, "Revised Notes on Gold and Silver Determination by Atomic Absoprtion Spectrophotometry," At. Absorpt. Newsl., Vol. 6, No. 2, p. 37.

32. Tsukahara, I., and Tanaka, M., 1980, "Determination of Silver in Copper and Lead Metals and Alloys and in Zinc and Selenium by Atomic Absorption Spectrometry, after Separation," Talanta, Vol. 27, p. 237.

33. Viets, J. G., 1978, "Determination of Silver, Bismuth, Cadmium, Copper, Lead, and Zinc in Geologic Materials by Atomic Absorption Spectrometry with Tricaprylmethylammonium Chloride," Anal. Chem., Vol. 50, p. 1100.

34. Walton, G., 1973, "A Method for the Determination of Silver in Ores and Mineral Products by Atomic-Absorption Spectroscopy," Analyst, Vol. 98, p. 335.

35. Wemyss, R. B. and Scott, R. H., 1978, "Simultaneous Determination of Platinum Group Metals and Gold in Ores and Related Plant Materials by Inductively Coupled Plasma-Optical Emission Spectrometry," Anal. Chem., Vol. 50, p. 1694.

36. Young, R. S., 1971, Chemical Analysis in Extractive Metallurgy, Chas. Griffin and Co. Ltd., London, England; Chapts. 16, 33.

Table 1. Status of Major Gold and Silver Analytical Methods

Method	Status
Fire Assay	Routine quantitative for most sample types; umpire; also used as collector for AAS and spectrographic methods.
Atomic Absorption Spectroscopy (AAS)	Routine quantitative for all sample types; probably most-used method.
Emission Spectrography	Quantitative for some applications; semiquantitative for others; more popular before AAS developed.
Inductively Coupled Plasma Spectroscopy	Just getting started; shows potential.
Spectrophotometric	Popular trace-level method in '50's and '60's; backup and confirmatory.
X-ray (XRF, EDX)	Special applications; usually qualitative
Neutron Activation	Best method for ultra-trace levels in small samples; preseparations make method slow.
Gravimetric	Formerly used for higher grade samples.
Titrimetric	Formerly used for higher grade samples; some high grade applications (e.g., anode sludges).
Electrometric (polarographic, amperometric and coulometric)	Special applications; good sensitivity and precision.

Table 2. Selected Atomic Absorption Methodology for the Determination of Gold and Silver

			Flame AAS		Flameless AAS		
Sample	Sample Size (g)	Sample Preparation	Minimum Concentration in Original Sample	Concentration in Final Solution (mg/L)[a]	Minimum Concentration in Original Sample	Concentration of Final Solution (mg/L)[b]	Reference[c]
Au/solids	1-10	Aq. Reg.[d]/ 1-6 MHCl	34 g/t[e]	0.3-2	---	---	M, SG
Ag/solids	1-2	Aq. Reg.[d]/ 1-6 MHCl	10 g/t	0.05-4	0.1 g/t	0.001-0.025	M, SG
Au/solids	1-10	Roast/ CN leach	3 g/t	0.3-2	0.3[f] g/t	0.05-1.5	M, SG
Ag/solids	1-2	HNO_3/ 5% NaCN- 0.2% NaOH	10 g/t	0.05-4	0.1 g/t	0.001-0.025	M, SG
Ag/solids	0.1-0.2	Aq. reg./ NH_4OH	10 g/t	0.05-4	0.1 g/t	0.001-0.025	P
Au/solids	1-10	Aq. reg./ xtn[d] or Aq. reg./ HCl/xtn	0.2 g/t	0.02-2	0.02[f] g/t	0.005-0.5	T3
Ag/solids	1-2	Aq. reg./ xtn[d] or Aq. reg./ HCl/xtn	0.6 g/t	0.01-1	0.001 g/t	0.001-0.025	T3
Au/solids	1-10	Aq. reg./ 1-4 MHCl/ resin sep'n	3 g/t	0.3-2	0.3[f] g/t	0.05-1.5	S
Au/CN^-	(g)	Dilution	0.5 mg/L	0.3-2	0.005 mg/L	0.05-1.5	M, SG
Ag/CN^-	(g)	Dilution	0.05 mg/L	0.05-2	0.005 mg/L	0.001-0.025	M, SG
Au/ aq. reg.	≤250	Dilution or evap/HCl/xtn	0.5 mg/L 0.01 mg/L	0.3-2 0.3-2	0.005 mg/L 0.001 mg/L	0.05-1.5 0.005-0.5	M, SG M, T3
Ag/aq.	≤250	HNO_3 to paste/5% NaCN-0.2% NaOH	0.01 mg/L	0.05-2	0.001 mg/L	0.001-0.025	M, SG
Au/CN^-	≤250	Acid/oxidant/ xtn	0.01 mg/L	0.02-2	0.0001 mg/L	0.005-0.5	M, SG
Au/other aq. sol'ns (thiourea, thiocyanates, etc.	≤250	Dilution or oxdn/vol redn/HCl/ xtn	0.5 mg/L 0.01 mg/L	0.3-2 0.01-2	0.005 mg/L 0.0001 mg/L	0.005-1.5 0.005-0.5	M, SG T3
Au/ organics	(g)	Dilute with solvent	0.5 mg/L	0.01-2	0.005 mg/L	0.005-0.5	M, SG, T3

a)ppm = mg/L divided by the specific gravity of the solution at ambient temperatures.
b)Values assume single injection of 25 L.
c)M = Mallett(1970); SG = Sen Gupta (1973); T3 refers to Table 3 references; S = Sukiman (1976); P = Perushottam, et al. (1972)
d)Aq. reg. = aqua regia; xtn = extraction.
e)1 g/t = 1 ppm = 0.0292 troy-oz/short-ton.
f)Based on 10 g sample and 50 ml final volume.
g)Needs only enough liquid for aspiration.

Table 3. Gold and Silver Extraction Methods

Solvent	Extractant	Aqueous Phases Reported	Metal Extracted	Selected References
MIBK	MIBK-Cl^-	Aqua Regia; 1-8 M HCl; 5% HCl+HBr+Br_2; Aqua Regia + permanganate + 3-8 M HCl	Au	Tindall (1965 and 1966), Hall (1979), Greaves (1963)
DIBK Toluene Xylene	DBS	Aqua Regia; 1-3 M HCl	Au	Rubeska, et al. (1977) Parkes and Murray-Smith (1979)
MIBK DIBK	TBP	Aqua Regia; 1-8 M HCl 8 M HNO_3	Au	Das and Bhattacharyya (1976)
MIBK DIBK Toluene	TOPO	HCl	Au	Das and Bhattacharyya (1976)
DIBK	TOA	NaCN, Aqua Regia	Au	Groenewald (1968, 1969), Mallett, et al. (1966)
DIBK MIBK	TOMA (or Aliquat 336)	NaCN	Au, Ag	Viets (1978), Groenewald (1968, 1969) Tsukahara (1980)
MIBK DIBK	5% TOPO	Aqua Regia + KI + ascorbic acid	Ag	Burke (1974)
MIBK DIBK	30% TOTP	Nitric acid	Ag	Bratzel, et al. (1972)
MIBK Butyl Acetate	0.5% DPT in Acetone	$HClO_4$, HNO_3, HF digest in 20% tartaric acid	Ag	Aruscavage and Campbell (1979)

DBS = dibutyl sulfide
DIBK = diisobutyl ketone
DPT = diphenyl thiourea
IAA = isoamyl alcohol
MIBK = methyl isobutyl ketone
TBP = tributyl phosphate
TOA = trioctyl amine
TOMA = tri-n-octyl methyl ammonium chloride
TOPO = tri-n-octyl phosphine oxide
TOTP = triisooctylthiophosphate

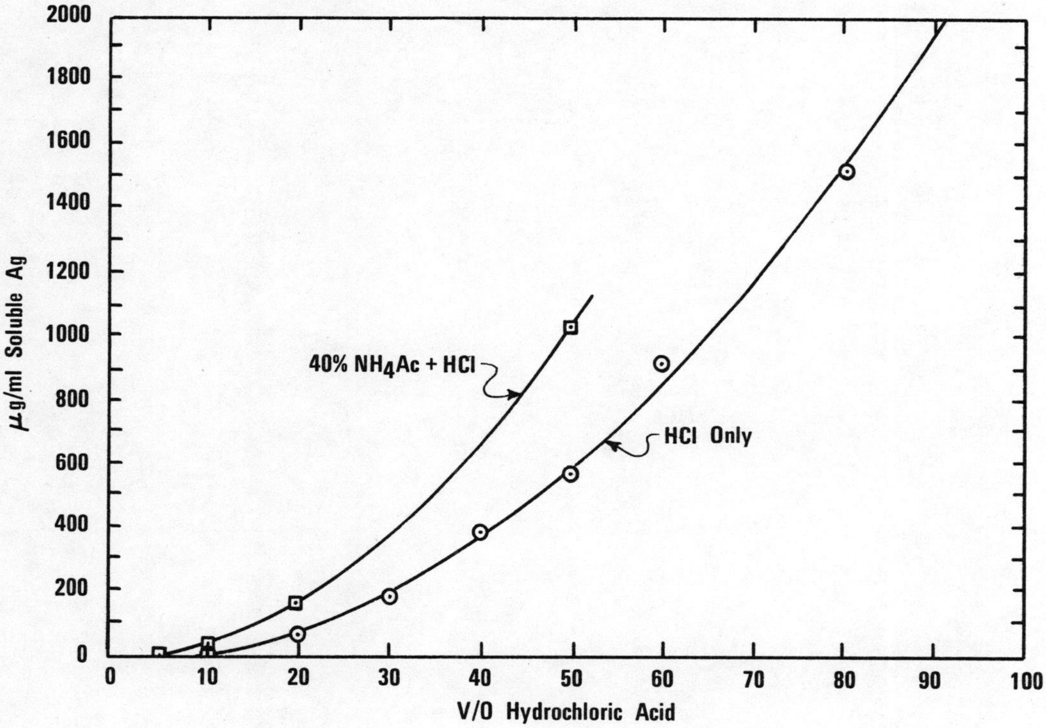

Figure 1. Solubility of silver in HCl–ammonium acetate solutions as determined experimentally

Chapter 13

REVIEW OF THE GOLD AND SILVER MARKETS

James L. Poole
Continental Bank
Chicago, Illinois 60693

ABSTRACT

The gold and silver markets from the late 1960's to the present are reviewed and the outlook for these metals in the 1980's is presented. Included in this discussion is a summary of the basic supply and demand balance for gold and silver since the mid–1960's, an explanation of the price formation for the metals, some of the relationships between the prices of these metals and selected economic parameters, and how these factors affect the outlook.

GOLD

Gold Supply. Gold supply is made up of three components: mine production, communist country sales and net official sales. The largest component of gold supply is mine production, which makes up about 55% of the total. Figure 1 and Table 1 show world mine production of gold which has been steady at 40 million troy ounces per year since 1974. The world's major producer is South Africa with about 60% of the total, the USSR is second with 22%, Canada is third with 4%, and the United States is fourth with 2-3% of total world production. The remaining production is made up from other countries, none of which is particularly major.

TABLE I

WORLD MINE PRODUCTION OF GOLD
(million ounces)

	S. Africa	U.S.S.R.*	Other	Total
1965	30.55	5.03	11.12	46.70
1966	30.88	5.37	10.85	47.10
1967	30.53	5.70	9.47	45.70
1968	31.17	6.25	8.89	46.42
1969	31.28	6.25	8.89	46.42
1970	31.39	6.50	8.86	47.52
1971	31.39	6.70	8.10	46.19
1972	29.25	6.90	7.97	44.12
1973	27.49	7.10	7.33	41.92
1974	24.39	7.30	6.97	38.66
1975	22.94	7.50	8.10	38.54
1976	22.94	7.70	8.33	38.97
1977	22.50	7.85	8.51	38.86
1978	22.65	8.00	8.49	39.14
1979	22.02	8.16	9.01	39.79
1980*	21.70	9.54	8.32	39.56

*estimated

Source: American Metal Market, J. Aron

The second source of gold supply, shown in Table II, is communist country sales, predominanatly from the USSR. In recent years, this amount has ranged annually from 11% to 29% of total gold supply. Sales by the USSR are dependent upon their foreign exchange requirements, which explains the variation in sales from year to year as both gold prices and the internal exchange requirements change.

The third source of gold supply, net official sales, shown in Table II, includes sales by the International Monetary Fund (IMF), government agencies such as the US Treasury, and other sources such as private holdings. It is important that both the IMF and US Treasury recently ceased their respective gold auctions. This could have profound effects on gold markets over the next several years.

Gold Demand. Gold demand or consumption is made up of two parts: Industrial demand which accounts for about 60-70% of the market; and speculative demand which makes up the remaining 30-40%. Industrial consumption primarily encompasses the requirements of the jewelry and electronics industries. The remainder is mostly dental and defense uses. Speculation or investment demand is essentially coins, gold for trading (speculation) or investment, and gold held for longer term appreciation (hoarding). Table III illustrates gold consumption by end use.

It is important to note that gold consumption by both the speculative and industrial markets varies considerably depending upon the price and outlook for gold. For example, the high price in early 1980 decreased gold consumption by the jewelry sector almost 40% from the same period in 1979; but the low prices in the first quarter of 1981 increased gold consumption by the industrial sector and central banks.

TABLE II

WORLD GOLD SUPPLY AND DEMAND
(million-troy ounces)

	1975	1976	1977	1978	1979	1980*
Supply						
Mine Production	30.6	31.1	31.1	31.2	31.2	30.7
From Communist Countries	4.8	13.2	12.9	11.6	7.8	3.5
Net Official Sales	0.3	1.9	8.6	11.6	17.0	(3.1)
Total Supply	35.7	46.2	52.6	56.0	56.0	31.1
Consumption						
Industrial Fabrication	23.5	38.5	40.7	41.6	35.0	13.8
Official Coins	7.8	5.9	4.4	8.3	9.0	7.5
Bullion Purchases**	4.4	1.8	7.5	6.1	12.0	9.8
Total Demand	35.7	46.2	52.6	56.0	56.0	31.1

* Estimate
** Residual estimate of supply available to meet investor demand for bullion.

Source: American Metal Market

Gold Prices. Gold price formation is a function of economic factors, supply and demand, which are dominated by the industrial market, and speculative factors. Figure 2 shows the price making influences for gold. The supply considerations are on the top half of the illustration and include the previously discussed supply factors. Mine production is the primary influence. The right side of Figure 2 is industrial consumption, and the left is net governmental demand. The lower right portion of the chart illustrates the factors which enter into the speculative portion of price formation. This area includes yields on competing assets, i.e., what is the interest rate and what are the risks involved?; inflation, value of the dollar, and a 'World Anxiety Factor'. This last factor includes both political and economic problems that arise from time to time. However, political uncertainty only causes a reaction in gold prices when the uncertainty has some economic ramifications.

To understand these influences on gold prices, it must be remembered that gold is considered a store of value. Thus, when the dollar devalues or inflation increases, the price of gold moves up. Or, when the general economic or political situation deteriorates, there is a flight away from currency and into gold. The converse is true when conditions stabilize.

To summarize, in gold price formation there are really two influences at work, one overlying the other. There are always the underlying industrial supply and demand relationships. Overlaying this are the speculative forces that fluctuate in intensity depending on the level of uncertainty, either economic or political.

TABLE III

WORLD GOLD CONSUMPTION BY USE
(million troy ounces)

Industrial	1975	1976	1977	1978	1979
Jewelry	16.6	30.0	32.1	32.2	26.5
Electronics	2.2	2.4	2.5	2.7	2.5
Other	4.6	6.0	6.3	6.7	6.0
Sub-Total	23.4	38.4	40.9	41.6	35.0
Investment					
Coins	7.8	5.9	4.4	9.0	9.0
Hoarding	0.0	5.7	2.2	2.6	2.5
Speculation	4.4	(3.9)	5.3	2.8	9.5
Sub-Total	12.2	7.7	11.9	14.4	21.0
TOTAL	35.7	46.1	52.8	56.0	56.0

Source: American Metal Market

SILVER

Silver Supply. Contrasting silver with gold, there are basic similarities and major differences. Silver supply is comprised of four sources: mine production, net official sales, scrap, and sales from India. The non-communist world produces about 350 million ounces of silver per year as shown in Figure 3 and Table IV. Mexico and the USSR vie annually for first place, although Mexico is expected to be the largest producer in the near future due to significant additions of new mine capacity. These two countries are followed by Peru, Canada, the US and Australia. A major difference between gold and silver mine production is that contrary to gold, no one country dominates in silver.

TABLE IV

MINE PRODUCTION OF SILVER
(Million troy ounces)

	Mexico	USSR	Peru	Canada	USA	Other	TOTAL
1966	42.0	33.0	32.8	32.8	43.7	82.4	266.7
1967	38.3	35.0	32.1	37.2	32.3	83.3	258.2
1968	40.0	35.0	36.4	45.0	32.7	85.3	258.2
1969	42.9	37.0	35.9	43.5	41.9	94.5	295.7
1970	42.8	38.0	39.8	44.3	45.0	98.7	308.6
1971	36.7	39.0	38.4	46.0	41.6	97.6	299.3
1972	37.5	40.0	40.2	44.8	37.2	105.5	305.2
1973	38.8	41.0	40.0	47.5	37.8	106.8	311.9
1974	37.5	42.0	38.3	42.8	33.8	106.7	301.1
1975	38.0	43.0	39.0	39.7	34.9	110.3	304.9
1976	42.6	44.0	37.0	41.2	34.3	115.5	314.6
1977	48.4	45.0	38.8	42.2	38.2	115.9	328.5
1978	50.8	50.5	38.6	40.9	38.9	123.6	343.3
1979	49.4	49.8	40.2	38.1	37.1	127.0	341.6
1980	52.1	50.0	40.3	42.5	39.8	133.9	358.6

Source: J. Aron, Silver Institute

The other sources of silver supply are shown in Table V. These include:

1. Government stocks, which are sales by various governments. These have dwindled considerably in recent years.

2. Scrap sales, which include everything from old silverware to waste from manufacturing processes.

3. Indian stocks, which are primarily privately held, ornamental or religious artifacts. This source supposedly decreased in February 1979 as a result of export restrictions placed on silver by the Indian government.

4. The last category or 'other' is mainly demonetized coin and imports from communist countries.

TABLE V

SUPPLY AND DEMAND FOR SILVER, EXCLUDING THE COMMUNIST BLOC COUNTRIES
(million ounces)

	1975	1976	1977	1978	1979
Supply					
Mine	242	247	267	265	271
Government Stocks	21	8	5	9	3
Scrap	73	76	80	86	81
Indian Stocks	66	70	41	45	26
Other	14	51	44	29	42
Total	416	452	437	434	433
Consumption					
Industrial	377	422	417	404	410
Coinage	39	30	20	30	23
Total	416	452	437	434	433

Source: Handy & Harman

Silver Demand. Silver demand, like gold, is made up of two facets: industrial and speculative consumption. Industrial consumption accounts for approximately 95% of all the silver used and the major consuming sectors are the photographic, jewerly, silverware and electronic industries. See Table VI. Speculative consumption accounts for the remaining. Coins and bullion make up the major portion of speculative demand, and here, as in gold, the consumption is very dependent on the current silver price and its outlook.

Silver Prices. Like gold, silver prices are a combination of both economic factors and speculative factors. Figure 4 shows the price making influences for silver. There are similarities between this chart and the one for gold, Figure 2; the top half has been changed to show the different sources of silver supply. Other than supply sources, the charts are the same because the demand forces that act on gold and silver are almost identical. That is, silver's price is a function of industrial supply and demand for the metal, and like gold, this base demand is overlain by a speculative demand for the metal. The important implication is, for speculative purposes, silver is a good substitute for gold.

PRICES AND ECONOMIC PARAMETERS

A discussion of gold and silver would be incomplete without a few comments on the relationships between gold and silver prices: to one another, to inflation, to oil prices, and the investment potential.

Figure 5 shows the tandem relationship between the price of gold and the price of silver from January 1970 through February 1981. It must be noted, however, that although gold and silver have similar markets they are different commodities. Therefore, events that might have an effect on the silver market, such as actions by the Hunt brothers in early 1980, may not affect the gold market and vice versa. Also, the gold market, in general, reacts much more quickly than the silver market. This was evident in March 1981 when the Russians threatened to invade Poland; the news of a possible invasion caused large movements in the price of gold while the silver price remained relatively stable.

TABLE VI
ESTIMATED USE OF SILVER IN THE USA, EUROPE AND JAPAN* IN 1980

Use	Million Ounces	Percent
Photography	125.2	37%
Jewelry/Silverware	64.1	19
Electrical/Electronic	70.7	21
Alloys & Solders	31.8	10
Coins	9.8	3
Other	32.8	10
Total	334.4	100%

* Accounts for 90% of the non Communist world's consumption.

Figure 6 illustrates the prices of gold and silver plotted against the annual percent change in inflation. The price of gold and the inflation factors are given in Tables VII and VIII. Of importance in Figure 6, is that after 1973 the prices and the inflation rate correlate quite well; decreases in the inflation rate lead to a lowering of gold prices and increases in the inflation rate lead to increases in the price of gold. This indicates that the prices of gold and silver are directly related to the inflation rate, once again maintaining the concept that gold and silver are long-term stores of value.

TABLE VII

ANNUAL AVERAGE GOLD PRICES

Year	U.S.Price $/oz.	U.S.Price $/oz. * (in 1967 dollars)
1966	35.00	36.01
1967	35.00	35.00
1968	39.26	37.68
1969	42.19	38.42
1970	36.41	31.31
1971	41.25	34.03
1972	58.61	46.78
1973	97.81	73.49
1974	159.73	108.29
1975	161.49	100.12
1976	124.77	73.18
1977	147.87	81.43
1978	193.55	98.95
1979	307.51	141.32
1980	613.03	248.59

* Deflated by the CPI (Consumer Price Index)
Source: American Metal Market

Figure 7 plots the prices of gold, silver and oil. Tables VII, IX and X detail the prices. This figure suggests that gold and silver prices lead oil prices, which implies that increased oil prices result from inflation and do not initially cause inflation. This can be seen on this graph in mid-1973 and in Table XI when gold prices began to escalate at a higher rate than oil prices. This phenomenon also occurred in 1978 when both the prices of gold and silver escalated at a rate higher than oil.

Investment Performance. The last comment with regard to prices is the investment performance of these commodities since 1967. An examination of Tables VII and IX shows the real or constant dollar price for both gold and silver have increased at just over 13% per annum during this 14 year period. This is the best return on any asset during that period, including real estate, with the possible exception of Chinese art. Also, since 1975, the real prices for gold and silver have increased 20% and 25% per year respectively. These returns speak not only to the speculative interest in gold and silver but also the investment potential of these metals. However, the potential of receiving similar returns in the future is not only a very controversial matter, but is subject to inconclusive debate of the type that always surrounds speculative matters.

TABLE VIII
CPI INFLATION FACTORS

Year	CPI Index	% Change from Previous Year
1966	97.2	2.75
1967	100.0	2.88
1968	104.1	4.10
1969	109.0	4.71
1970	116.3	6.70
1971	121.2	4.21
1972	125.3	3.38
1973	133.1	6.23
1974	147.5	10.82
1975	161.3	9.36
1976	170.5	5.70
1977	181.6	6.51
1978	195.6	7.71
1979	217.6	11.25
1980	246.6	13.33

Source: U.S. Department of Commerce

TABLE IX
ANNUAL AVERAGE SILVER PRICES

Year	U.S.Price $/tr oz	U.S.Price $/tr oz (in 1967 dollars)*
1966	1.29	1.33
1967	1.55	1.55
1968	2.15	2.06
1969	2.14	1.95
1970	1.77	1.59
1971	1.55	1.28
1972	1.68	1.34
1973	2.55	1.92
1974	4.71	3.20
1975	4.42	2.74
1976	4.35	2.55
1977	4.62	2.54
1978	5.40	2.76
1979	11.09	5.10
1980	20.63	8.37

* Deflated by the CPI
Source: American Metal Market

OUTLOOK

Forecasting for gold and silver is very difficult because of the speculative nature of the commodities. There are, however, certain things now occurring in these markets, which impact the outlook issues for gold and silver.

In regard to gold:

1. Mine production from South Africa is decreasing; there will be potentially less sales from the USSR, as they get more foreign exchange from less gold sold because of higher prices; and sales from the IMF and U.S. Treasury have all but ceased.

2. Higher prices will reduce industrial demand, but higher prices increase hoarding; possibly netting out the effects of decreased industrial demand and thereby keeping total demand strong.

3. Based on the assumptions of decreasing supply and relatively stable demand, a tight market for gold in the near to medium term is possible.

In regard to silver:

1. Increasing silver production from mines and sales from large private holdings indicate a surplus of silver.

2. High prices will reduce demand through permanent substitution and lower sales, but these effects may be netted out by increases in hoarding.

3. These assumptions, especially increasing supply, suggest a soft market for silver in the short to medium term.

The following comment about future gold and silver prices is in order: In stating that the price of gold and silver will remain near current levels for the next five years, it is implied that 1) there will not be a major international political or economic crisis, and 2) that inflation will abate significantly during that period of time. A scenario that is difficult to comprehend in light of events of the last 10 years!

Figures 1 through 7 appear on the following pages.

TABLE X
OIL PRICES

Year	Annual Average Price for Marker Crude U.S.$/bbl	Annual Average Price for Marker Crude* U.S.$/bbl. (1967 dollars)
1966	1.80	1.85
1967	1.80	1.80
1968	1.80	1.73
1969	1.80	1.64
1970	1.80	1.55
1971	2.18	1.80
1972	2.48	1.98
1973	3.31	2.99
1974	11.58	7.85
1975	12.38	7.68
1976	12.38	7.26
1977	13.33	7.34
1978	13.66	6.98
1979	18.53	8.52
1980	28.63	11.61

* Deflated by the CPI
Source: OPEC Annual Statistical Bulletin

TABLE XI
COMPARISON OF % CHANGE
from
PREVIOUS YEAR FOR:
GOLD, SILVER AND OIL PRICES and
THE INFLATION RATE

	% Change From Previous Year			
Year	Gold Price	Silver Price	Oil Price	CPI
1966	0.0	0.0	0.0	2.75
1967	0.0	20.16	0.0	2.88
1968	12.17	38.71	0.0	4.10
1969	7.46	(0.47)	0.0	4.71
1970	(13.70)	(17.29)	0.0	6.70
1971	13.29	(12.43)	21.11	4.21
1972	42.08	8.39	13.76	3.38
1973	66.88	51.79	33.47	6.23
1974	63.31	84.71	249.85	10.82
1975	1.10	(6.16)	6.91	9.36
1976	(22.74)	(1.58)	7.37	5.70
1977	18.51	6.21	7.67	6.51
1978	30.89	16.88	2.48	7.71
1979	58.88	105.37	35.65	11.25
1980	99.35	86.02	54.51	13.33

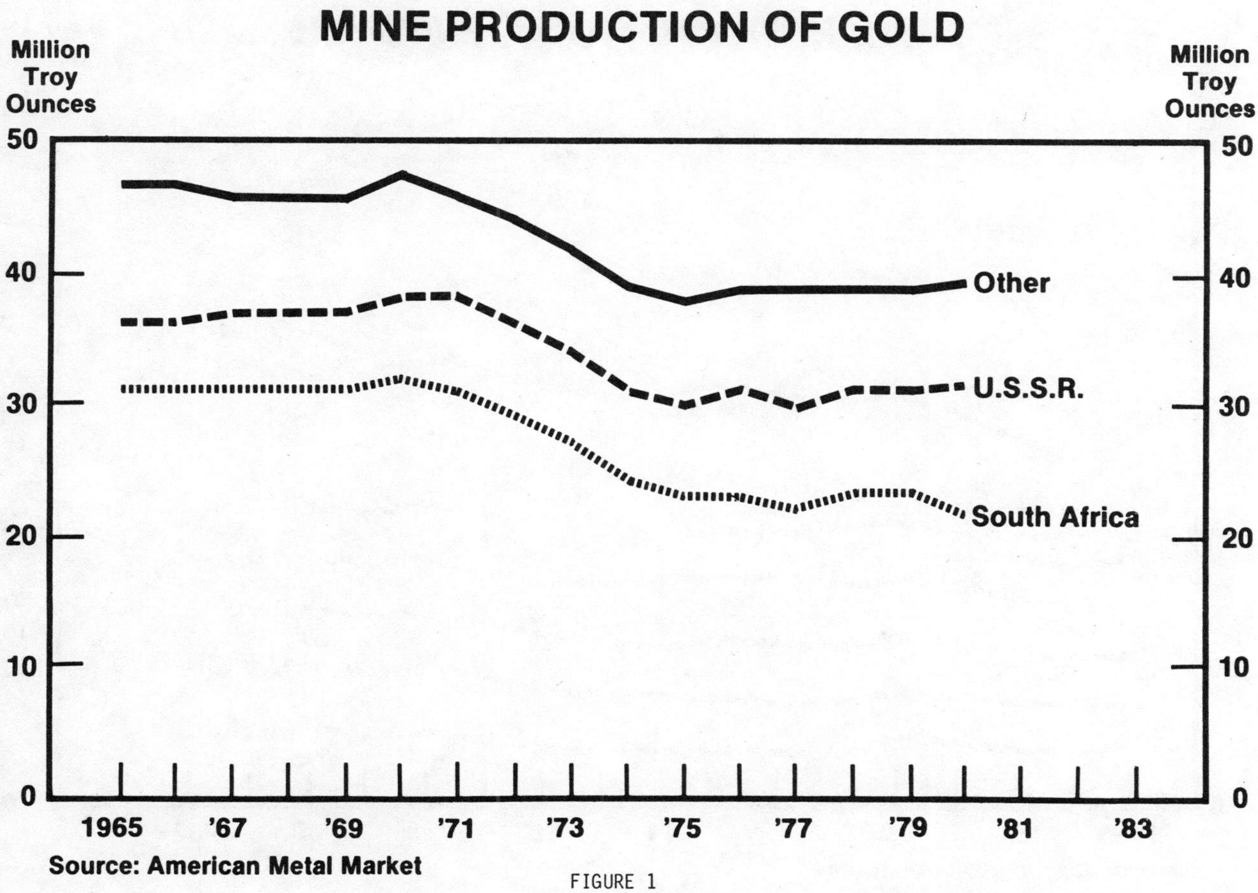

FIGURE 1

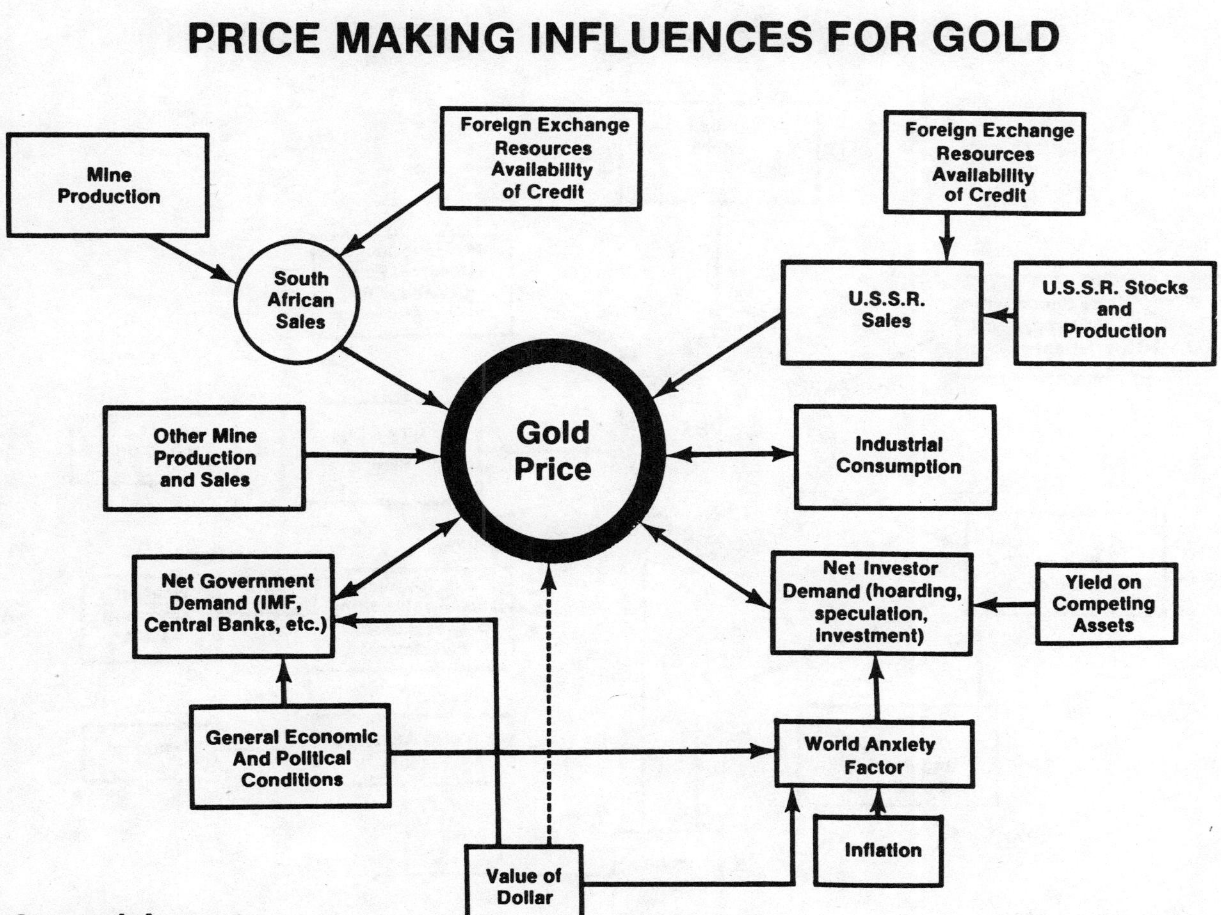

FIGURE 2

MINE PRODUCTION OF SILVER

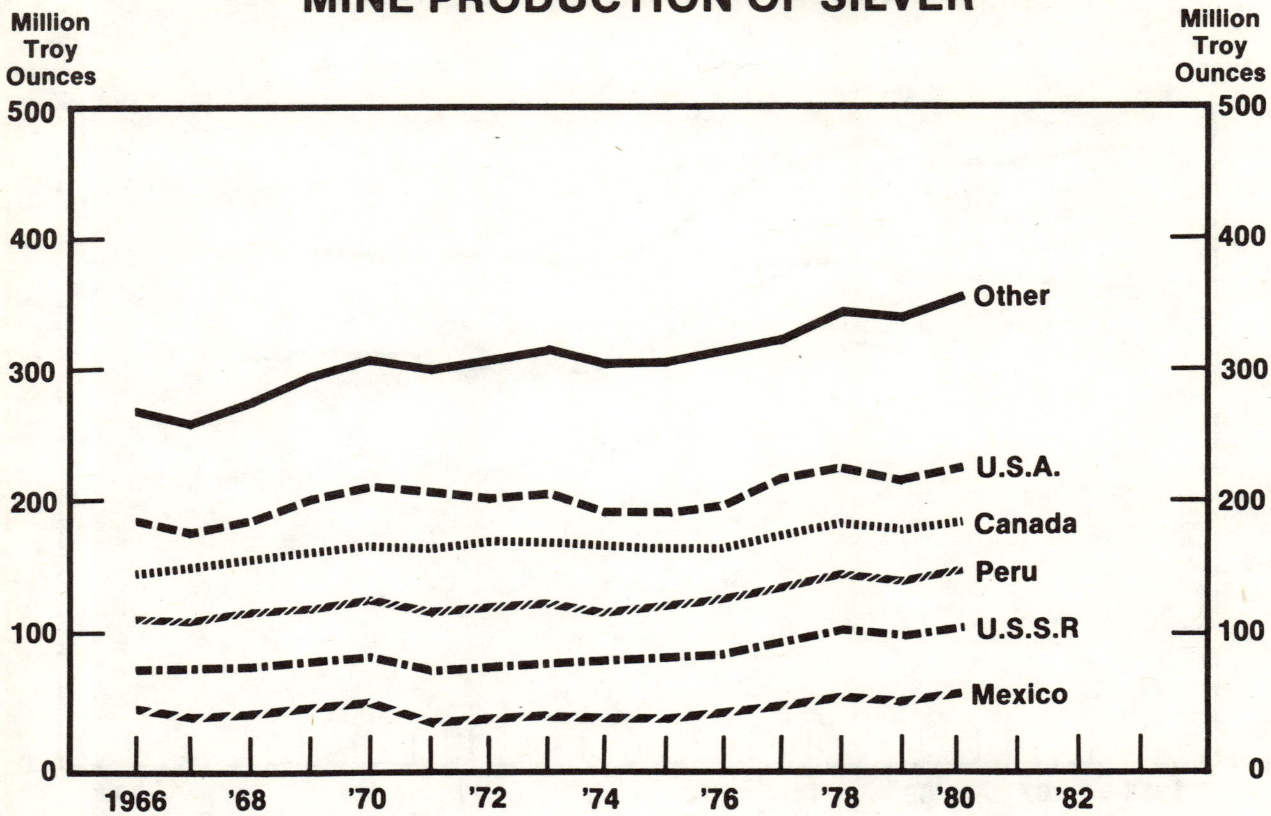

Source: J. Aron, Silver Institute

FIGURE 3

PRICE MAKING INFLUENCES FOR SILVER

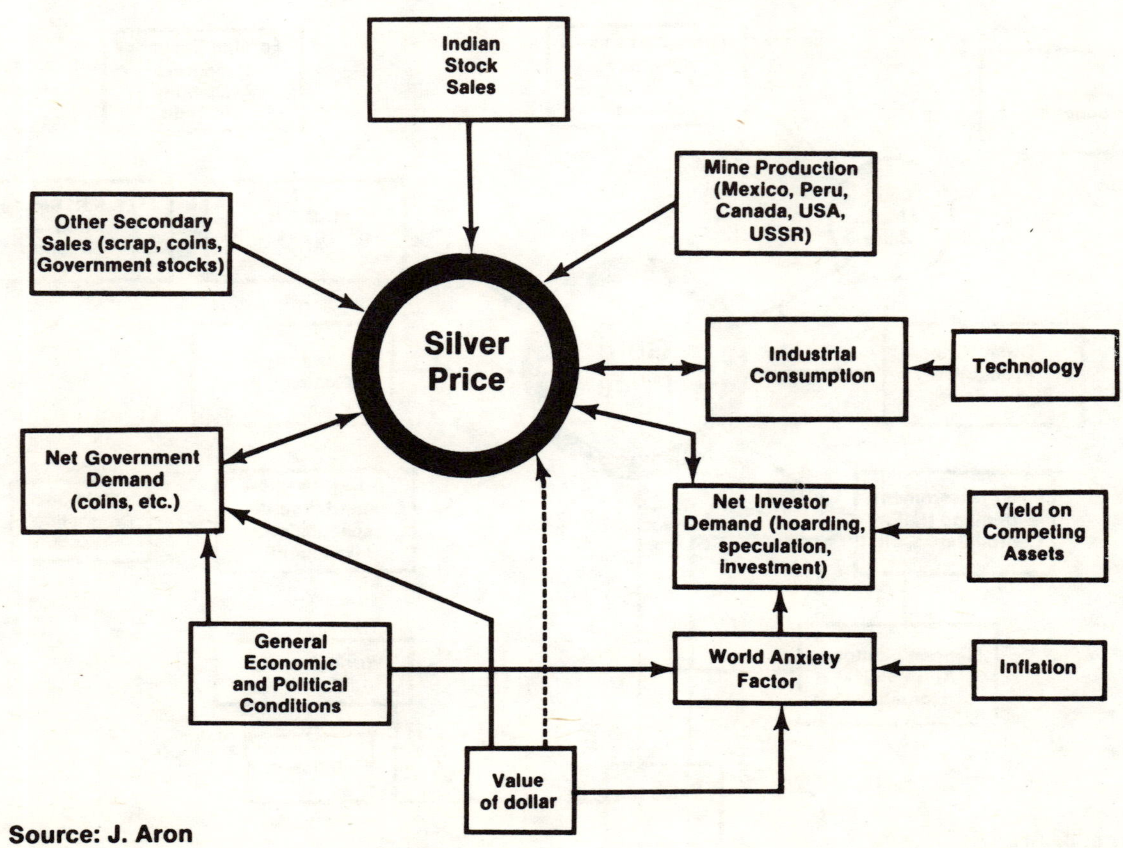

FIGURE 4

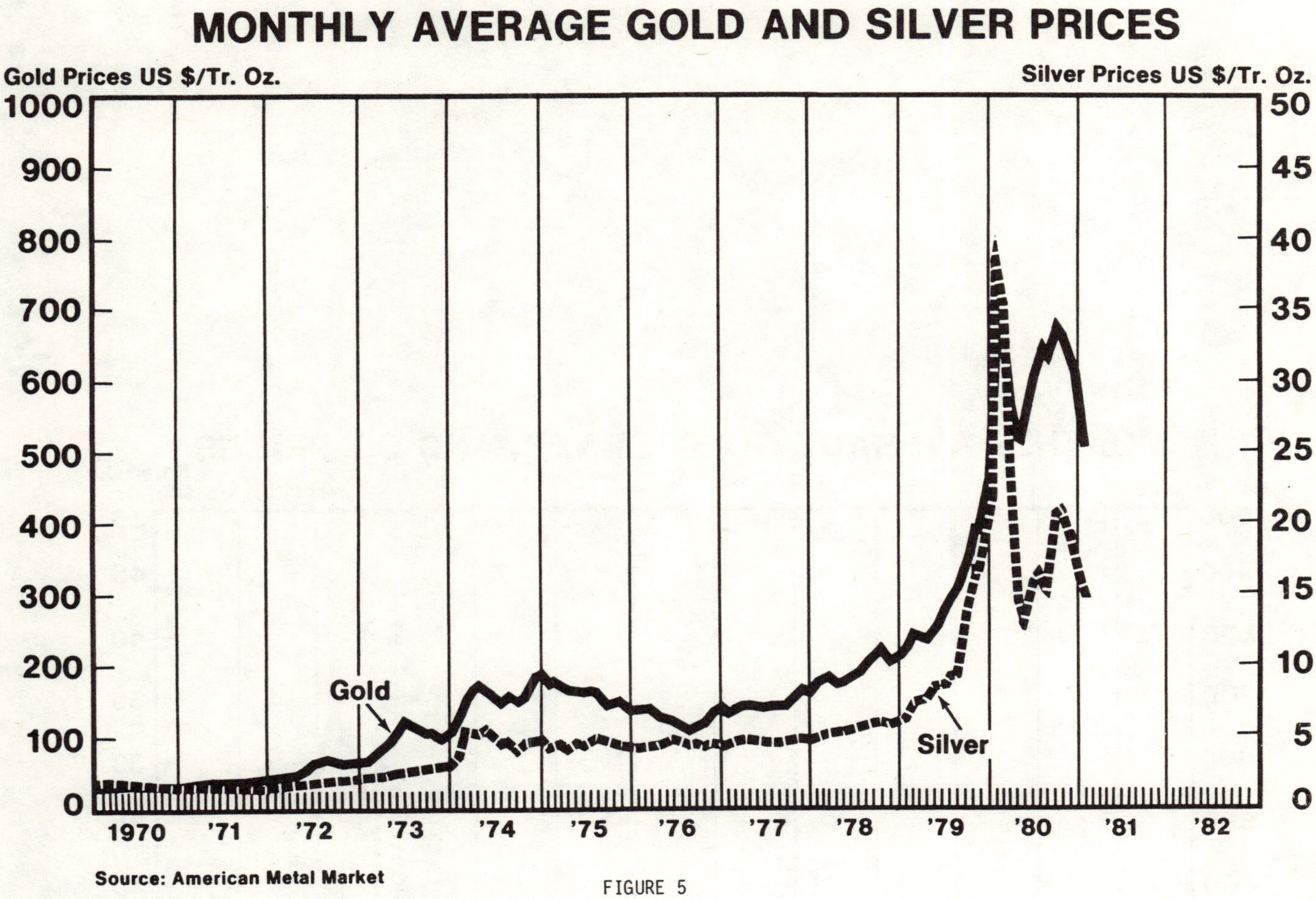

FIGURE 5

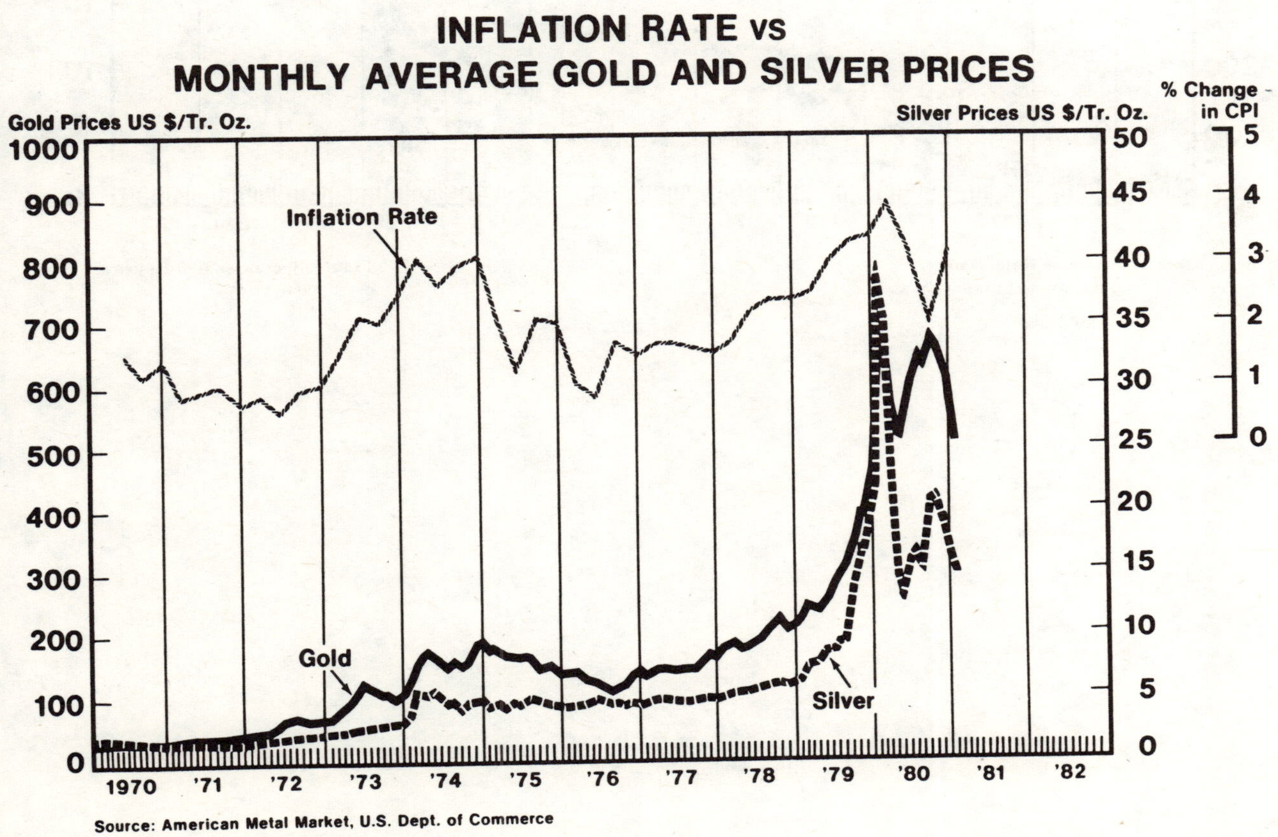

FIGURE 6

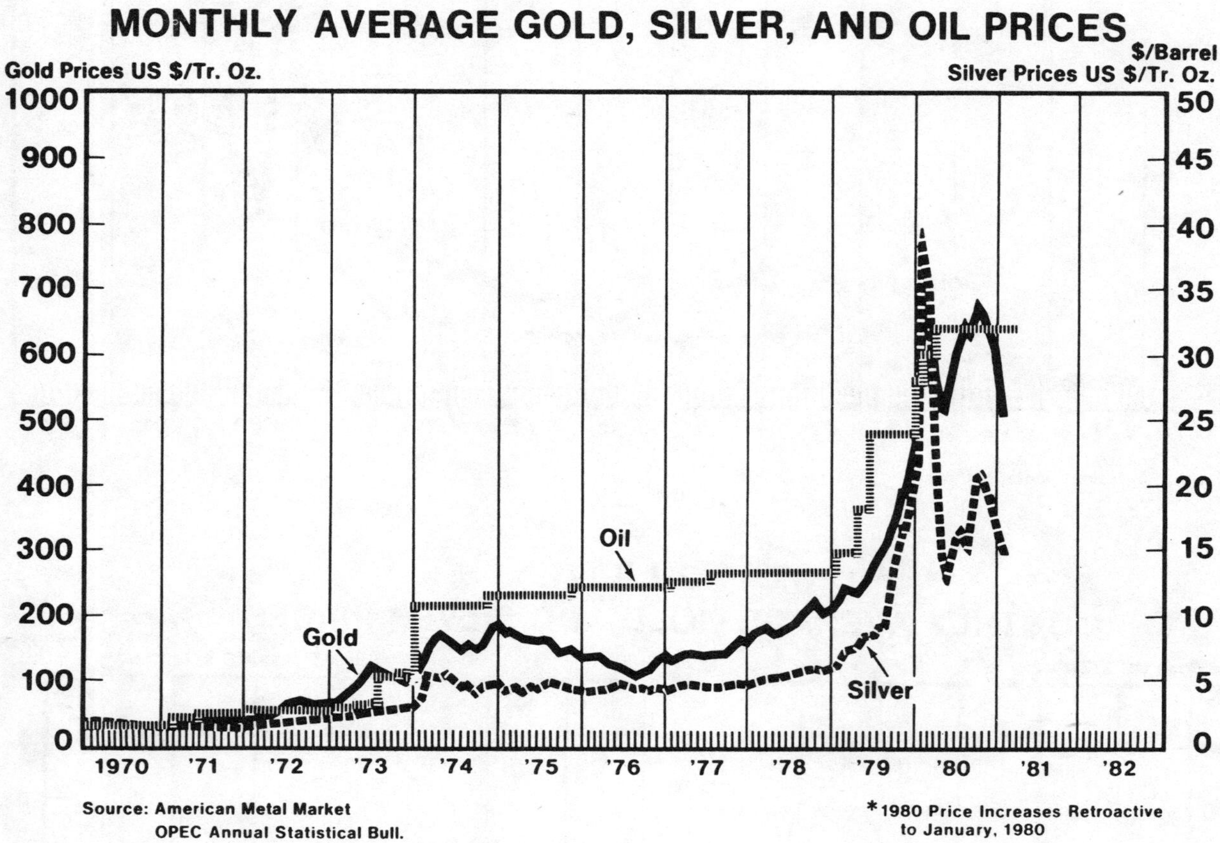
MONTHLY AVERAGE GOLD, SILVER, AND OIL PRICES
Gold Prices US $/Tr. Oz.
$/Barrel
Silver Prices US $/Tr. Oz.
1000
900
800
700
600
500
400
300
200
100
0
50
45
40
35
30
25
20
15
10
5
0
Oil
Gold
Silver
1970
'71
'72
'73
'74
'75
'76
'77
'78
'79
'80
'81
'82
Source: American Metal Market
OPEC Annual Statistical Bull.
*1980 Price Increases Retroactive
to January, 1980

FIGURE 7

Chapter 14

THE ECONOMICS OF SILVER AND THE MINER

Hugh Douglas
Hugh Douglas & Company Ltd.
534 Pacific Avenue
San Francisco, CA 94133

ABSTRACT

In the past two years recent increases in silver prices over levels that prevailed for many years have changed supply/demand patterns. On the supply side, a silver mining boom is in progress in the United States and elsewhere. On the demand side, traditional uses of silver by individuals are dropping while industry is conserving. These changes will alter the supply/demand balance and bring about more moderate rises in silver prices, which in turn will impact the mining industry.

INTRODUCTION

Miners and explorationists historically have been driven to dig and search for a metal by one single factor: price. There are other noneconomic factors such as government subsidies (in fact a price subsidy), the need for some consumers to have captive supplies, and the desire for most geologists just to practice their art of mine finding.

When the price of a metal or mineral is high exploration budgets are available to look for that metal. If the price of the metal goes high enough, exploration syndicates and promoters join the mine finding and development activities. Ultimately such activity leads to discoveries, thence to new capacity with the help of financing from large mining houses, and then over capacity and falling prices for the quested metal.

In the long run the consumer has never been without a supply of metals.

Last year was extraordinary when the price of silver rose to over $48 per troy ounce on 21 January 1980 and fell back precipitously to $11.10 by 27 March. Today, silver is selling for a little over $13 per ounce. Will silver ever again see such lofty prices, and at the very least, will it sell for prices that warrant the expectations that the mining industry, geologists, mining engineers and others, are placing on silver?

This question will be answered, not by presenting a detailed supply/demand balance of the world silver industry, but by highlighting statistics on silver and showing a 130 year price series in 1980 constant dollars, that is, dollars or money without the effects of inflation. From this review one can have some perspective on the future course of the silver industry and prices.

SUPPLY

For the past 30 years and more the supply of newly mined silver in the world and in the United States has been inadequate to meet demand. The difference came from hoards and stocks.

From a low 135 million ounces in 1946, world silver production steadily increased and doubled to a peak of 270 million ounces in 1979. Last year production was 255 million ounces; the decline due to strikes in the base metal industries, and mining to a lower grade in silver mines.

A characteristic of world silver production is that since 1945 silver has been produced increasingly as a byproduct of base metal mining whereas historically there had been a large percentage of silver mining from deposits containing primarily silver. Thus, the output of silver has recently been geared to the production of base metals and this output has little connection with the price of silver.

The recent and present turmoil in silver markets and prices reflect actions taken by the US Treasury many years ago. In the United States the coinage laws of 1834 and 1837 adjusted the gold:silver ratio to 16:1. This was done by setting the metal content of the dollar at 23.22 grains of gold or 371.25 grains of silver with a fineness of 900 (90%). Gold was fixed at a value of $20.67 per ounce, and by legal definition, silver had a monetary value of $1.293 per ounce. The standards remained unchanged until 1934 when the price of gold was raised to $35 per ounce, thus reducing the gold content of the dollar. To maintain the historic ratio of the two metals, the amount of silver should have also been reduced to give a monetary value of $2.19 per ounce. Legislation to effect this change was considered but Congress instead authorized the Treasury to purchase silver until its stocks, valued at $1.293 per ounce, should equal one-third of the Treasury gold stock valued at $35 per ounce and to issue silver certificates equal to the value of its holdings.

This policy attempted to give silver backing to some of the paper currency; however, it fixed the price of silver so that additional primary supplies were constrained. With rising industrial growth the demand for silver grew at rates faster than would have been likely had the price of the metal been set by market forces.

Thus, with rising silver consumption the stocks built up by the Treasury for fifteen years were exhausted in 1961 and silver sales were suspended. In 1965 the Treasury ceased minting silver coins. With the Treasury stocks no longer available, the additional supplies for the silver markets have come from stocks, melting of old coins and from sales by traditional hoarders in India and Pakistan.

In a report prepared last year, 'A Guide to the US Gold and Silver Mining Industries'(1) a boom taking place in silver mining and production in the United States over the next several years was forecast. The research indicated that in the period between 1983 and 1985, according to announced plans by the mining industry, silver production in the US would increase to about 70 million ounces, or over twice last year's output of 31 million ounces. A survey of the mining literature shows that similar activities are taking place world wide.

The structure of the US mining industry will also change over the next several years, see Table 1. As the new mines come on stream, the share of production of the current top 11 silver producers will decline from 46% of the total silver output in 1979 to 34%. There will be more mines and smaller ones.

Table 1

STRUCTURE OF THE U.S. SILVER MINING INDUSTRY

1979 AND 1980-83

(Millions of Troy Ounces)

	1979		1980-83	
	Amount	Percent of total	Amount	Percent of total
11 major companies	20.1	46%	24.6	34%
All others	23.7	54	48.7	66
Total	43.8	100%	73.3	100%

Source: Hugh Douglas & Company Ltd, San Francisco Mining Associates

The Silver Institute, which has as its members most of the silver mining companies in the world, states that world production by 1983 could rise to over 411 million ounces from an estimated 355 million ounces in 1980. This is an increase of 16%.

DEMAND

Consumption of silver last year fell in every major industrial country in the world. In the US, silver demand declined to 119 million ounces in 1980 from 157 million ounces in 1979; world demand dropped from 420 million ounces to 340 million ounces. The major end use markets, silverware and photography, showed the biggest declines.

The increasing price has worked to push the demand for silver down and brought out new technology to replace or reduce the amount of silver used in industrial applications. Nowhere is this new technology more apparent than in photography. In the past six months the following developments were announced.

- Ilford of England, a Ciba-Geigy subsidiary, has developed a process that recovers nearly all the silver from developing black and white film. That advance had already been achieved with color films. The new film is capable of high speeds and very fine grained images.
- Kodak recently announced a process that uses less silver.

In flatware, heavy, good looking stainless steel is fast replacing silver, not only for price reasons, but also because it is easier to maintain. For the medium income family the additional insurance now charged for having silverware in the home is also a constraint.

Elsewhere in the electronics industry vigorous efforts are underway to both reduce the amount of silver used and to recover waste silver.

SUPPLY AND DEMAND

United States production and consumption of silver for 1979 and 1980 are shown in Table 2. In 1980 supply increased over 1979 by nearly 30 million ounces, primarily from old scrap. However, because of lower imports and higher exports, net supply last year actually declined. With lower demand, the deficit was about the same. This year, new mines and steady output from the copper industry should increase new mine production, demand will continue to fall, and silver supply and demand are expected to be in balance.

Table 2

UNITED STATES
PRODUCTION AND CONSUMPTION OF SILVER

1979 AND 1980*

(Millions of Troy Ounces)

	1979	1980
Ores and concentrates	38.1	31.0
Coins	4.5	5.0
Old scrap	36.0	70.0
Subtotal	78.6	106.0
Imports	92.3	76.3
Gross supply	170.9	182.3
Less Exports	35.6	82.5
Net supply	135.3	99.8
Industrial demand	157.3	119.7
To (from) stocks	(22.0)	(19.9)

*Preliminary

Source: Handy & Harman,
Hugh Douglas & Company Ltd

On a world basis, supply/demand in 1980 is shown on Table 3. New mine production declined by 15 million ounces from the prior year. Salvage increased by 30 million ounces. Private holders of bullion sold silver in 1979 whereas they acquired silver in 1980. Overall, supplies in 1980 increased by nearly 100 million ounces from the 1979 total of 472.7 million ounces.

Table 3

WORLD SILVER SUPPLY/DEMAND BALANCE
1980
(Millions of Troy Ounces)

Supply		Demand	
New mine production	255.0	Industrial	340.2
US Treasury	0.1	Coinage	15.7
India/Pakistan	41.7	Subtotal	355.9
Coinage melting	55.0	Additions to private bullion stocks	
Government stocks	5.2		
Salvage	121.5		122.6
Total	478.5	Total	478.5

Source: Handy & Harman

AN HISTORICAL OVERVIEW

Historical trends are not necessarily a portent of the future. The silver industry is behaving in a classic response to prices—supply and demand are coming into balance and this year they are likely to be in balance. The caveat raised most often is that with continued depreciation of the world's currencies through inflation, silver will be a refuge, a store of value, as is gold. But one tends to look at the past twenty years as the yardstick for inflation and not the next ten years.

In major industrialized countries, governments are re-evaluating their monetary policies. Unlimited demand for nearly every major commodity is tapering off. In the light of these developments is silver still a hedge against inflation? The primary demand for silver will be industrial and this demand is declining. At the same time newly mined silver is rising. The market system at work may make silver a poor hedge against inflation.

Figure 1 shows the long term trend of silver prices stated in constant 1980 dollars. (The GNP Implicit Price Deflator was used to convert current dollars into constant dollars.) Also shown is world production since 1905 and world consumption for the last ten years.

Prior to 1980, the all time high for silver in constant 1980 dollars was in 1851 at $13.83 per ounce. From that time the price has been decreasing until the 1970s when rapid increase in demand, slowly rising supplies, and speculative hoarding and trading pushed the price up. If one were to look at the historical trend and eliminate last year's aberation, one would conclude that the settlement price for silver is between $10 and $15 per ounce.

This price is adequate to pay for the byproduct recovery of silver. If silver is the primary metal, then for an underground mine and mill with costs of $75 per ton of ore, one would need at least 5 ounces of silver per ton to recover costs, and about 9 ounces per ton to yield a 15% return on the investment. Silver values could be less if there are other metals to be recovered.

It is prudent to so some homework before committing hundreds of thousands of dollars for silver exploration ventures and millions more to open up a mine.

(1) 'A Guide to the US Gold and Silver Mining Industries', Hugh Douglas & Co. Ltd. and San Francisco Mining Associates, San Francisco, CA, 1980.

FIGURE 1 WORLD CONSUMPTION, PRODUCTION AND PRICE OF SILVER, 1850–1981

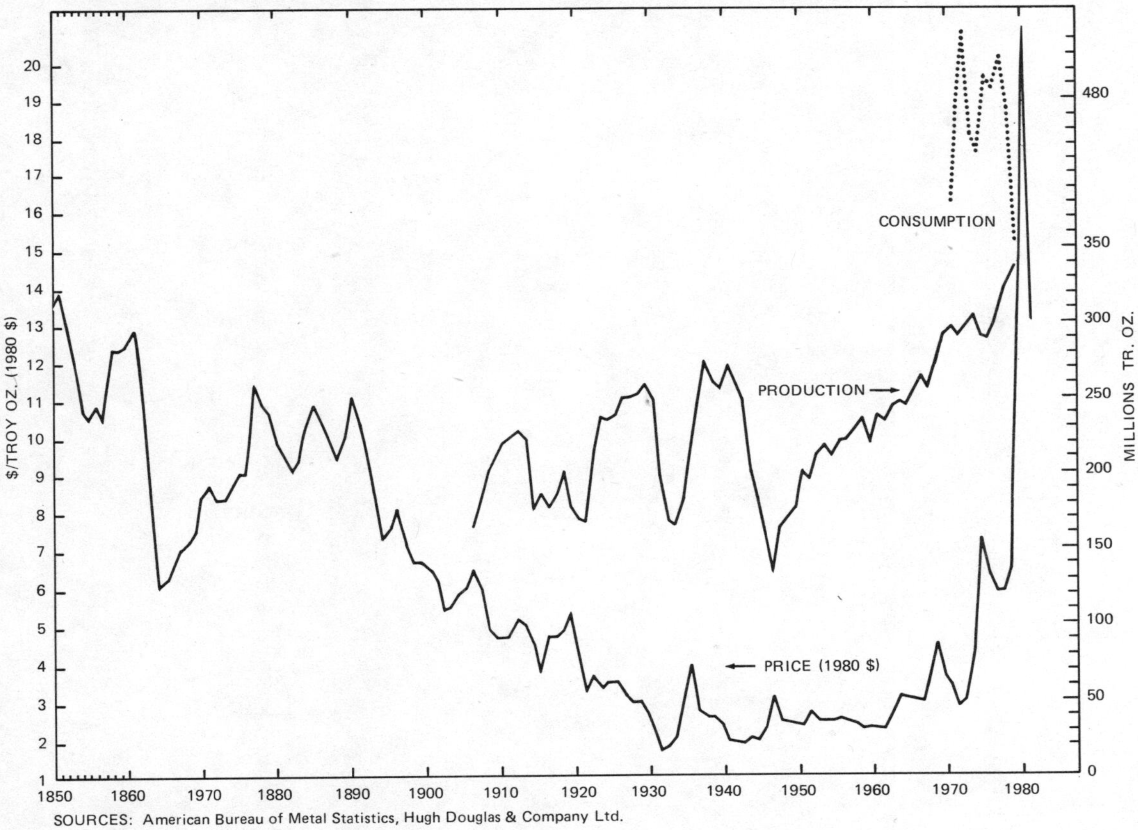

SOURCES: American Bureau of Metal Statistics, Hugh Douglas & Company Ltd.

Chapter 15

PRECIOUS METALS ECONOMICS

Paul Sarnoff
Rudolf Wolff Commodity Brokers Inc.
295 Madison Avenue
New York, NY 10017

ABSTRACT

The economics of precious metals -- depending upon who is involved -- could be the economics of hope, of fear, indeed of profit and perhaps survival. For those investors seeking capital gains in a puzzling economy, precious metals involve the economics of hope. Those who are convinced that currency soon won't be worth the paper it's printed on turn to precious metals because of the economics of fear. Gold and silver have been a storehouse of wealth since biblical times. The economics of fear during the past 500 years on the Continent has caused almost every European family to make sure a private stockpile or hoard of gold, silver or coins is at hand in case the paper currency of their country undergoes the kind of disaster that came to the German Mark after World War I. Understandably, for mining companies involved in the production of precious metals, the elements of operations and management are directly concerned with the economics of profit -- and perhaps survival.

To the American miner the price of precious metals in the marketplace is a matter of daily concern. After decades of selling silver to the US Government at 64.64 cents an ounce, producing silver mines suddenly became faced with prices breaking $5, $10 and $20 per ounce, and approaching $50 an ounce during the period 1978–1981. What should the miner expect one year in the future? Five years?

In 1979 the average daily cash price of silver was \$11.09 per troy ounce; in 1980 the average daily cash price of the white metal rose to \$20.63. Silver companies pursued expansion by sinking deeper shafts, or working abandoned properties whose ore had become unprofitable in previous years because of low prices. When silver prices escalated to new highs toward 1980, Sunshine Mines even recycled their parking lot because it contained about 2 ounces of silver per tonne. There is ore in Hecla's Lucky Friday mine which runs 87 ounces of silver per tonne and higher. The wisdom of sinking a new shaft to get at more of that ore hinges on whether or not the price of silver will remain above \$7.50 an ounce, the estimated base cost of retrieval.

IMPACT OF PRICE ON MINERS

In South Africa, a nation that currently supplies 73% of the free world gold, the price of gold each day in London, New York, Chicago, Hong-Kong and other world centers is of vital concern to both the government and its mining operators. For the past five years the daily average price of gold in the case market in New York has been:

YEAR	AVERAGE PRICE OF GOLD (per troy ounce)
1980	\$612.51
1979	307.59
1978	193.44
1977	148.32
1976	124.96

A higher gold price is almost the equivalent of better profits for the mining companies.

Unlike mining operations in the United States, South African gold mines are required, in terms of their leases from the State, to mine according to the average value of their ore reserve. Robin A. Plumbridge, Chairman, Gold Fields of South Africa explains, '. . . In some mines, a relatively small downward or upward movement in the pay limit renders substantial tonnages of low grade ore payable or unpayable and this can have a profound effect on the total ore reserve tonnage and value. ...We now review our pay limits monthly so that our mines are able to mine approximately to the average value of the ore reserves from month to month. ... if the pay limit shows a declining tendency, the crew from a high grade working place which has been exhausted will be move to a new working place of lower value'.

Referring to the wild gyrations in the gold price during 1980, Mr. Plumbridge revealed, 'By using gold prices and individual mine's working costs averaged over a period of six months, we believe that a smoothed resulting pay limit provides a basis for allowing adjustments on each mine to be made with a minimum of disruption; and we are optimisitc that our mines will now be able to respond more sensitively to the changing circumstances which we see as an inevitable part of gold mining operations in the future. ...'

It is almost an iron-clad corollary that if the price of gold goes up, things will improve in South Africa, in the mining sector at least; whilst if the price of gold declines drastically, the profit picture will radically change. This price gyration affects more than corporate profits, because improved prices and profits allow better pay and living conditions for the workers.

PRICE AND PRICE RATIOS

Gold, silver and platinum are mainly traded and quoted in dollars, with minor markets in other currencies. Gold, like oil, is internationally quoted, primarily in dollars; but silver has a major alternative market in the rings at the London Metal Exchange, where it is quoted in pounds sterling. The primary cash market in platinum is also made in London, in sterling, but is a dealer market and not an exchange traded metal.

The prices of the precious metals differ for many reasons, not all of them logical. Exchange rate differentials will affect prices, as will changes in parity or the relationships between the prices of each of the metals. For example, at this time it takes about 40 ounces of silver to buy one ounce of gold; but for almost 100 years the relationship in the world free market stood at 32 to one. When Columbus discovered America it took 10 ounces of silver to buy one ounce of gold; when the USA went on a bimetallic system in 1896, the relationship officially set by government fiat was 16 to one. Figure 1 depicts the changes in the gold/silver ratio from 1500 AD to 1980.

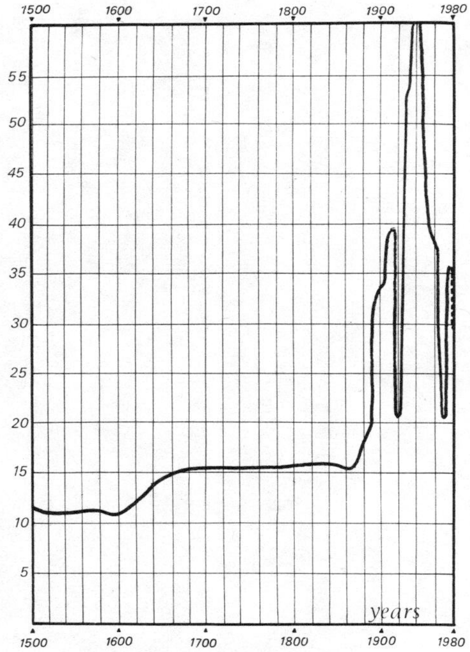

Source: Societe Miniere et Metallurgique de Penarroya

It is worth noting that the average ratio for four hundred years stood around 15 to one. On the basis of the historical record gold is currently greatly overpriced in relation to silver. Mr. Gerin-Jean, President of Penarroya, believes the relationships between gold and silver will continue to change but that '. . .one ounce of silver should approach 1/20th the price of one ounce of gold'. If this relation turned out to be true and gold stood at \$500, then silver would be \$25 per ounce, or twice the value prevailing in early 1981.

The price of gold ought to be higher than the price of silver because of the relative scarcity of gold. About ten times as much silver is mined worldwide each year as gold. A simplistic but possibly fair evaluation of the silver price on the basis of relative scarcity may be one to ten, rather than one to forty as it is today. Bunker Hunt publicly stated in 1979 when the silver price was rising, that he believed the time would come when silver would be a five to one ratio, that is, it would take only five ounces of silver to buy one ounce of gold. At the time of this prediction the ratio had moved down from the 32 to one of the previous year to about 27 to one. In the third week of January 1980 the ratio sank toward the 15 to one level. Immediately after the silver price collapse in March 1980, the ratio ran up over 40 to one and has more or less gravitated within ten percent of that figure since.

Gold is primarily a monetary metal, with the central banks of the world holding the major protion of the visible gold stocks. However, silver and platinum are primarily industrial metals. The principal use of silver is photography; the principal uses of platinum involve catalytical applications in automobile antipollution devices and the petroleum industy. The current depressed state of the American automobile industry has adversely affected the platinum price to the point where it is trading $20 or more an ounce under the gold price. This may be only a temporary aberration. For the past five years the price of platinum on a daily basis has been substanially higher than the daily gold price. For example, the daily average cash platinum price in 1980 was $677 per ounce whilst the average gold price was $612. During January 1980 gold rose to a zenith of $850 an ounce but later in the year platinum rose to over $1150, demonstrating the upside potential for this often neglected metal.

METAL MARKETS

In olden — and perhaps golden — times, before futures markets, bullion dealers made the price of precious metals. The advent of progress and the broadening of investor and speculator interest in forward contracts resulted in exchanges which traded contracts in precious metals. At this time precious metals are traded on the following commodity exchanges:

METAL	EXCHANGES
Gold	COMEX - NY Commodity Exchange
	Chicago Board of Trade
	Chicago Mercantile's IMM Division
	MidAmerica
	Winnipeg
	New York Mercantile
	Hong-Kong
	Singapore
	Australia
Silver	COMEX
	London Metal Exchange
	Chicago Board of Trade
	MidAmerica
Platinum	New York Mercantile Exchange

In addition, the London Gold Market, which is the physical market that issues the gold fix twice a day, is planning a London Gold Futures Exchange together with the Ringmembers of the London Metal Exchange. This gold futures exchange in London will forge the last link in the global gold markets, and will provide access to listed trading of gold contracts for both speculators (people who want to profit from price changes) and hedgers (mining and consuming firms that use futures markets for price protection), twenty-four hours a day.

The futures exchanges that trade in precious metals contracts generally are not geared for physical delivery of metal. However, the purchaser of buy contracts on any precious metals exchange can get delivery if the contracts held in his or her account are kept open during the delivery month. Thus a holder of a contract to purchase 100 ounces of COMEX 995 gold at $500 an ounce for April 1981 delivery can expect a notice that the metal was going to be delivered to his or her account as soon as April made its debut. Under normal circumstances the holder of such April gold buy contracts will liquidate them by offset (sell the buy contracts in order to cancel them and their open and existing position) long before the delivery month. The holders of the major portion of the open long interest in a precious metal future may choose to stand for delivery and thus precipitate a market crisis. The shorts (hedgers or speculators who have sold short in anticipation that the price will drop when the longs liquidate their contracts) caught in such a crisis will often start to scream that they are being squeezed. Such a crisis condition received widespread publicity in the press during 1979 and 1980 when it happened in the silver market. (1)

Precious metals futures exchanges provide the means whereby producers and consumers of precious metals can shift their risks to speculators. Thus the producer can go short (sell for delivery) many months ahead if the price on the exchange is high enough to make such commitment worthwhile, and the consumer can go long (buy for delivery) many months ahead to insure against a steep price rise and other market risks.

In years past the prices of precious metals, like those for base metals, remained fairly stable; but in recent years the volatility of precious metal prices has been exacerbated by speculative efforts external to normal supply/-demand forces. This has resulted in growing attention by both consumers and producers on strategies which resort to the futures markets as hedges or 'insurance' against market-related losses.

MONEY AND PRECIOUS METALS

During the past decade the United States policy has been to get the gold out of the money. At one time paper money was redeemable in both gold and silver. The gold certificates suffered a demise around 1933, whilst the silver ones disappeared after 1967. In place of paper backed by a precious metal, the government now issues similar looking dollars--with the exception that the legend on them is simply 'Federal Reserve Note'.

The Federal government does have precious metals held in storage: about 264-million ounces of gold and 139.5-million ounces of silver. From May 1978 until the practice was halted in the fall of 1979, the Treasury sold over 7-million ounces of gold at an average price below $250 an ounce. This undervalued disposal pales in significance compared to the economics of selling silver from a stockpile that stood over 2-billion ounces in the late sixties to the current stockpile of 139.5-million ounces. The price of the silver which was sold or exchanged for paper ranged from $1.29 an ounce to $4.25 -- with an average of $1.69 -- a monumental transfer of wealth from the Treasury to the consumers and hoarders of the metal.

It is worthwhile speculating about the economics of again backing US money with precious metal. During the Reagan presidential campaign in the last quarter of 1980, consideration was given to the effect that gold might again have monetary significance. Research shows that when the US entered upon a bimetallic monetary system in 1896 there was $1 in gold in the US Treasury for every $24 in paper in circulation in the US. The amount of paper in circulation in 1979 both in the United States and outside the country was equivalent to $1 in gold for every $17,000 in paper.

Since last year our offshore supply of paper dollars has grown to over $900-billion. There appears little chance in the foreseeable future for US currency to be backed by gold in the disciplined manner of the Swiss Franc or the Deutsches Mark. But who can tell? It may not be time to back circulating paper with gold or silver.

However, the Treasury could follow the lead set up by Sunshine Mining which floated millions of dollars worth of gold-indexed bonds in two separate issues during 1980. In both cases the interest on the long term bonds was substantially lower than the market rate because the bondholders were provided with long-term calls on silver. A separate issue by Refinemet was similarly secured by calls on gold, resulting in an interest rate yielding less than 4%, far below the market rate. These innovative offerings suggest that the issuance of long-term, low interest bonds backed by Government silver and gold might be a feasible means to reduce the growing debt-service burden on the trillion dollar debt of the Federal government.

SUPPLY OF PRECIOUS METALS

The visible, aboveground stocks of gold are enough to satsify fifty years of demand, even if every gold mine closed tomorrow. However, the same is not true in the case of silver nor for platinum. The economics of silver dictate that the emerging nations becoming used to the comforts and advances of Western civilization -- computers, TV sets, mirrors, photovoltaic cells, etc. -- will place a severe strain on the aboveground stocks of silver.

For example, the Soviet Union has become increasingly westernized in certain areas that entail increasing silver demand. Whereas the United States uses about 3/4 ounce of silver per person per year, the Soviets currently use about 1/2 ounce per person per year. Since the Soviet nation encompasses more than 300-million people the existing demand is about 150-million silver ounces per year. In 1980 the Soviet Union led the world in silver output at 49-million ounces, suggesting a shortfall of about 100-million silver ounces. The United States also has a shortfall of about 100-million ounces which is made up by recycled scrap, coin melt, and purchases from Canada, Mexico and Peru. Thus the world silver supply and demand are in close, but good, balance.

Severe shortages can be anticipated in four or five years if China, with a population of one billion people, becomes as westernized in its silver needs as the Soviet Union is today. The Chinese would then require 1/2 ounce of silver per capita; it would generate national demand that will exceed the 1985 expected world silver production of 424-million ounces. The result may be an invisible trade war, or perhaps even a visible one for the needed silver resources. Such an explosive demand will pressure the price of silver upward; and it can be expected that the gold/silver price ratio at that time will be 10 to one instead of the current 40 to one.

The current trading of platinum substantially below the price of gold is bizarre. How long this will last is a question involving the anticipated recovery of the American automobile industry, and the possible purchase by the US government of platinum metal for the national stockpile. There is more than ten times as much gold in the form of ore reserves and in metal stocks available around the free world than there is platinum.

Burgeoning demand from an auto industry recovery will pervade the physical market and raise the free world platinum price. Until several years ago the Soviet Union was a significant supplier of platinum group metals to the free world, and it actually produces half the platinum group metals in the entire world. In 1979 the Soviet Union suddenly stopped selling platinum to the free world and began to stockpile the metal, leaving the US essentially 100% dependent upon South Africa. While silver will someday explode upward in price because of strains from the demand sector, it is likely that someday platinum will leap upwards in price because of lack of supply -- especially if the United States is serious about stockpiling strategic metals to match the Soviet preparedness for a 'conventional' war.

PRICE EXPECTATIONS

Where will the prices of precious metals go if the nations of the world continue on the current paths of rising inflation, rising cost of power and supplies, and political uncertainty?

If inflation continues at current double digit rates inside the United States and the price of oil keeps rising, then precious metals have only one way to go in the long-term, and that is up. On the way to $1000 gold, $100 silver and $1500 platinum there will be periods when prices will decline drastically; but that's what makes commodity markets so interesting when they are not tampered with by bureaucratic rules. Market volatility will be based on supply/demand conditions in the physical markets, and the supply/demand situation in the orders to buy and or sell futures contracts. The markets will rise and fall to meet requirements. Simply, precious metals prices are made by the persons most anxious to do business. Today's seller anxiety will depress prices whereas the anxiety of buyers will put the prices up. The long-run expectations for the precious metals suggest anxious buyers and rising prices.

REFERENCES

(1) Sarnoff, Paul, 1980, 'The Silver Bulls', Arlington House, Westport, CT, 190p.

INDEX

A

B

C

D

E

F

G

H

I

J

K

L

M

N

O

P

Q

R

S

T

U

V

W

X

Z